Fundamentals of Incompressible Fluid Flow

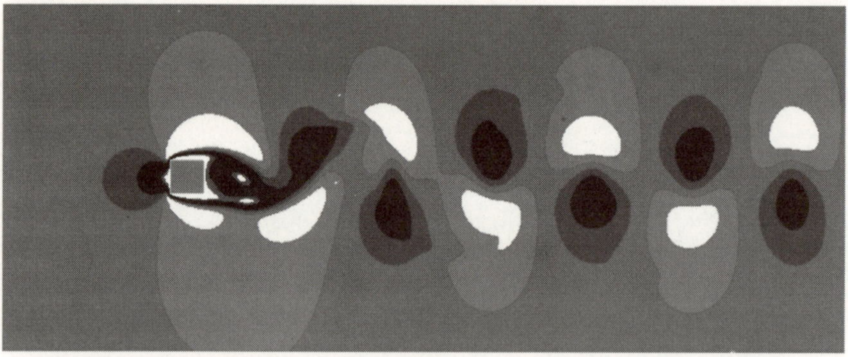

Cover illustration: Incompressible flow around a square cylinder (shown in gray) at a Reynolds number of 200 simulated using the lattice Boltzmann method. The von Karman vortex street can be seen clearly.

Fundamentals of Incompressible Fluid Flow

V. Babu
Professor
Department of Mechanical Engineering
Indian Institute of Technology,
Madras, India

Taylor & Francis
Taylor & Francis Group
Boca Raton London New York

CRC is an imprint of the Taylor & Francis Group,
an informa business

Ane Books Pvt. Ltd.

Fundamentals of Incompressible Fluid Flow
V. Babu

First Published in 2010

Ane Books Pvt. Ltd.

4821 Parwana Bhawan, 1st Floor
24 Ansari Road, Darya Ganj, New Delhi -110 002, India
Tel: +91 (011) 2327 6843-44, 2324 6385
Fax: +91 (011) 2327 6863
e-mail: anebooks@vsnl.net
Website: www.anebooks.com

For

CRC Press
Taylor & Francis Group
6000 Broken Sound Parkway, NW, Suite 300
Boca Raton, FL 33487 U.S.A.
Tel : 561 998 2541
Fax : 561 997 7249 or 561 998 2559
Web : www.taylorandfrancis.com

For distribution in rest of the world other than the Indian sub-continent

ISBN : 978 1 43985 4631

British Library Cataloguing in Publication Data
A catalogue record for this book is available from the British Library

Printed at Thomson Press, Delhi

Dedicated to my guide Prof. Seppo Korpela and his wife Terttu Korpela for their love and affection

Preface

I am pleased to bring out this book on Incompressible Fluid Flow based on my lecture notes of the postgraduate courses that I teach at IIT Madras on this subject. These classes are usually attended by final year B.Tech students, M.Tech, M.S and Ph.D students who have had a preliminary exposure to the subject matter earlier. I have assumed the same about the readers, while writing this book. This has allowed me to present concepts such as viscosity and Reynolds number much earlier than is usual. A broad range of basic concepts is presented in Chapter 2. The incompressible Navier-Stokes equations are derived in Chapter 3 and the mathematical nature of the solutions to these equations is discussed in Chapter 4. In this context, the notion of singular perturbation solutions - outer and inner, is introduced. Chapter 5 deals with the inviscid (outer) solutions, while the boundary layer (inner) solutions are derived in Chapter 6. Separation of the boundary layer, its consequences and drag are also discussed in detail. Analytical solutions, both parallel and creeping flow solutions, are presented in Chapter 7. Turbulent flows are discussed in Chapters 8, 9 and 10. The nature of turbulent flows and the importance of the turbulent mean flow are discussed in Chapter 8. Chapters 9 and 10 build on the latter idea in the context of internal and external flows respectively. Ideas of practical importance such as drag reduction in such flows are also presented.

The concepts and ideas discussed here are not new, but have merely been presented in a different manner arising from my experience in teaching them. Although there are many books on Fluid Mechanics, it is my hope that the readers might find the arrangement and the discussion of the topics in this book to be refreshingly different and insightful. If there are any errors or if you have any suggestions for improving the exposition of any topic, please feel free to communicate them to me via e-mail (vbabu@iitm.ac.in). As with my other books, my intention was not to write a book that was exam-oriented but one that will help students improve their understanding of the subject matter. To a large extent, the examples and exercise problems are drawn from practical applications to enable the student to appreciate the usefulness of the concepts discussed herein in such situations. I have suggested a list of books at the end (in alphabetical order) for the students to consult. These books also have a large collection of exercise problems that the students can (and should) practice.

Some parts of this book were written when I was at the Ohio State University on long leave. I wish to thank the Chairman of the Department of Mechanical Engineering, Prof. K. Srinivasan for the hospitality and support that I received during my stay there. On the personal side, I wish to thank Dr. Prasad Mokashi and his wife Janhavi for treating me as a member of their family and making my stay so pleasant and memorable.

As always, I would like to thank my teachers Sri. S. Sundaresan, Prof. R. Bodonyi, Prof. M. Foster and Prof. T. Scheick, who, as teachers, inspired me to a great extent. One person to whom I owe the most is Prof. S. Korpela of the Ohio State University. I learnt from him not only how to be a good academic researcher, but, more importantly, how to be a good human being as well. The kindness and affection that he and his wife, Tepa showed during my stay in Columbus last year was incredible. I have made an attempt to show them my gratitude by dedicating this book to them.

I cannot adequately express in words my gratitude for the love, affection, support and encouragement that I have received from my wife Chitra and son Aravindh over the years. I am fortunate indeed.

I wish to express my heartfelt gratitude to my grandfather Sri. V. Gopalan and my parents who endeavored so much to give me a good education. They have given a lot to me but received very little in return.

Finally, I would like to thank my students S. Somasundaram, V. G. Ramanathan and Dr. P. S. Tide for their help in working out the examples and exercise problems and my wife Chitra for proof reading the manuscript.

V. Babu

Contents

Chapter 1

Introduction

Fluid flow is a common phenomenon that we experience every day. Flow of water from an open tap or through a pipe and flow of air through a fan are a few examples. Most of these flows belong to a class of flows known as incompressible flows, which is the focus of this book. Our objective is to gain an understanding of the dynamics of such flows by determining the velocity and the pressure field. Once these are determined, quantities of engineering interest such as drag, pressure drop, flow rate etc., can be calculated. Assuming a flow to be incompressible simplifies the governing equations and hence the complexity of the solution to a large extent. Since all fluids are compressible to some extent or the other, it is important, to understand the limitations of this assumption and also be able to determine its validity for the flow situation under consideration. We turn to this next.

1.1 Compressibility of Fluids

The compressibility of a fluid is defined as

$$\tau = -\frac{1}{v}\frac{\partial v}{\partial P}, \tag{1.1}$$

where v is the specific volume and P is the pressure. The change in pressure corresponding to a given change in specific volume, will, of course, depend upon the compression process. That is, for a given change in specific volume, the change in pressure will be different between an isothermal and an adiabatic compression process.

The definition of compressibility actually comes from thermodynamics. Since the specific volume $v = v(T, P)$, we can write $dv = \left(\frac{\partial v}{\partial P}\right)_T dP + \left(\frac{\partial v}{\partial T}\right)_P dT$. Here, T is the temperature of the fluid. From the first term, we can define the isothermal compressibility as $-\frac{1}{v}\left(\frac{\partial v}{\partial P}\right)_T$, and, from the second term, we can define the coefficient of volume expansion as $\frac{1}{v}\left(\frac{\partial v}{\partial T}\right)_P$. The second term represents the change in specific volume (or equivalently density) due to a change in temperature. For example, when a gas is heated at constant pressure, the density decreases and the specific volume increases. This change can be large, as is the case in most combustion equipment, without necessarily having any implications on the compressibility of the fluid.

If the above equation is written in terms of the density ρ, we get

$$\tau = -\frac{1}{\rho}\frac{\partial \rho}{\partial P}, \tag{1.2}$$

The isothermal compressibility of water and air under standard atmospheric conditions are $5 \times 10^{-10}\ m^2/N$ and $10^{-5}\ m^2/N$. Thus, water (in liquid phase) can be treated as an incompressible fluid in all applications. On the contrary, it would seem that, air, with a compressibility that is five orders of magnitude higher, has to be treated as a compressible fluid in all applications. Fortunately, this is not true when flow is involved.

1.2 Compressible and Incompressible Flows

It is well known from high school physics that sound (pressure waves) propagates in any medium with speed which depends on the bulk compressibility. The less compressible the medium, the higher the speed of sound. Thus, speed of sound is a convenient reference speed, when flow is involved. Speed of sound in air under normal atmospheric conditions is 330 m/s. The implications of this when there is flow are as follows. Let us say that we are considering the flow of air around an automobile travelling at 120 km/hr (about 33 m/s). This speed is 1/10th of the speed of sound. In other words, compared with this speed, sound waves travel

10 times faster. Since the speed of sound *appears* to be high compared with the characteristic velocity in the flow field, the medium behaves as though it were incompressible. As the flow velocity becomes comparable to the speed of sound, compressibility effects become more prominent. In reality, the speed of sound itself can vary from one point to another in the flow field and so the velocity at each point has to be compared with the speed of sound at that point. This ratio is called the Mach number, after Ernst Mach who made pioneering contributions in the study of the propagation of sound waves. Thus, the Mach number at a point in the flow can be written as

$$M = \frac{V}{a}\,, \tag{1.3}$$

where V is the velocity magnitude at any point and a is the speed of sound at that point.

We can develop a quantitative criterion to give us an idea about the importance of compressibility effects in the flow by using simple scaling arguments as follows. From Bernoulli's equation for steady flow, it follows that $\Delta P \sim \rho V_{char}^2$, where V_{char} is the characteristic speed. The speed of sound $a = \sqrt{\Delta P/\Delta \rho}$, wherein ΔP and $\Delta \rho$ correspond to an isentropic process. Thus,

$$\frac{\Delta \rho}{\rho} = \frac{1}{\rho}\frac{\Delta \rho}{\Delta P}\Delta P = \frac{V_{char}^2}{a^2} = M^2\,. \tag{1.4}$$

On the other hand, upon rewriting Eqn. 1.2 for an isentropic process, we get

$$\frac{\Delta \rho}{\rho} = \tau_{isentropic}\Delta P\,.$$

Comparison of these two equations shows clearly that, in the presence of a flow, density changes are proportional to the square of the Mach number†. It is customary to assume that the flow is essentially incompressible if the

†This is true for steady flows only. For unsteady flows, density changes are proportional to the Mach number

change in density is less than 10% of the mean value. It thus follows that compressibility effects are significant only when the Mach number exceeds 0.3.

Let us explore this further through an example. Consider a hot plate at a temperature of 320 K placed horizontally in still air at 300 K. With time, the air above the hot plate becomes warmer and hence lighter and rises up due to buoyancy force. The colder and denser air farther away from the hot plate sinks down to replace the warmer air. This sets up a natural convection current in the vicinity of the hot plate. It is intuitively obvious that the velocity of the air will be quite small in such a situation and hence the flow can be safely said to be incompressible. However, since the driving mechanism in this case is the buoyancy force arising from the variation in density, it would be worthwhile to determine whether the density variation is small enough compared to the mean value. Since the change in density arises from a change in temperature, we can write

$$\rho = \rho_0 \left[1 + \frac{1}{\rho_0}\frac{\partial \rho}{\partial T}(T - T_0)\right] = \rho_0\left[1 - \beta(T - T_0)\right] ,$$

where $\beta = -\frac{1}{\rho}\left(\frac{\partial \rho}{\partial T}\right)_P$ is the co-efficient of volumetric expansion defined earlier†. Since air obeys the ideal gas equation of state, $P = \rho RT$, where R is the particular gas constant, it is easy to show that $\beta = -1/T$. Hence

$$\frac{\Delta\rho}{\rho_0} = \frac{\rho - \rho_0}{\rho_0} = -\frac{T - T_0}{T_0} .$$

For the example considered here, $|\Delta\rho/\rho_0| = (320 - 300)/300 \approx 7\%$ and hence the flow can be assumed to be incompressible, confirming our intuition. However, a temperature difference of 30 K or higher will make

† Alternatively, we can start from the definition of the coefficient of volumetric expansion,

$$\frac{d\rho}{\rho} = -\beta dT ,$$

and after integration, show that $\rho = \rho_0 e^{-\beta(T - T_0)}$. For small values of $T - T_0$, this can be written as $\rho \approx \rho_0\left[1 - \beta(T - T_0)\right]$.

the density change significant enough for the incompressibility assumption to be invalid.

It must be clear that the conditions for incompressibility that are given above, namely, that the Mach number be less than 0.3 or that the density variation be less than 10% of the mean value, are *ad hoc*. In the next chapter, the mathematical condition for considering a flow to be incompressible is given.

1.3 Laminar and turbulent flows

Flows can also be classified as being laminar or turbulent. In the former case, the fluid motion occurs in layers (*laminae*) and the motion is described quite well by closed form analytical solutions. In the latter case, the flow field exhibits fluctuations both in space and time and the motion cannot be described in a simple manner as for laminar flows. Statistical techniques have to be used to extract the required information from the flow field. Unfortunately, almost all of the flows seen in nature are turbulent rather than laminar. Laminar flows can be seen in real life only under extremely controlled circumstances since these flows are highly unstable and the tiniest of disturbances will make them undergo transition to a turbulent flow. The situation, however, is not as bleak as it seems, since it is still possible to get engineering estimates of the quantities of interest even for fully turbulent flows. These aspects are explored in detail in the following chapters.

Chapter 2

Basic Concepts in Incompressible Flows

In this chapter, a few basic concepts in the context of incompressible flows are discussed. These concepts offer insights into the flow by helping to depict as well as interpret the flow field. They are also used later in the derivation of the governing equations.

2.1 Definition of a fluid

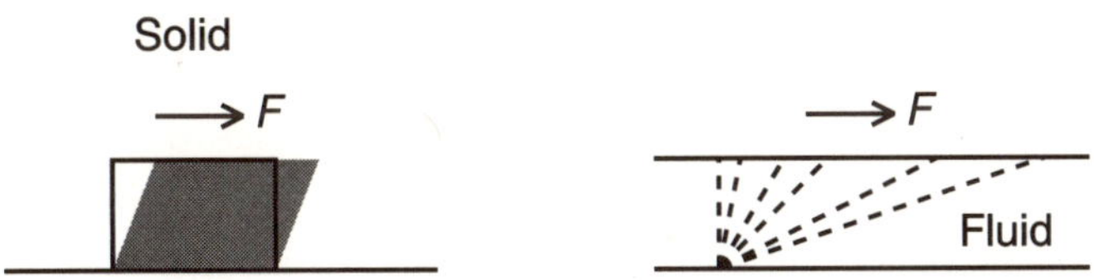

Fig. 2.1: Behavior of a solid and a fluid under an imposed shear force F

A fluid is defined as a substance that cannot sustain shear stress at rest. Figure 2.1 illustrates the behavior of a solid and a fluid under an imposed shear force F. The solid cube undergoes an angular deformation as shown in the figure. The fluid contained between two plates on the other hand, yields completely and begins to *flow* as evidenced by the successive orientations of an initially vertical (dashed) line. It is important to note that the scenario sketched in Fig. 2.1 for the solid will hold as long as the applied shear stress is well below the value at which the solid will yield completely. However, for the fluid, any non-zero value for the shear stress will result in fluid flow.

2.2 Steady and unsteady flows

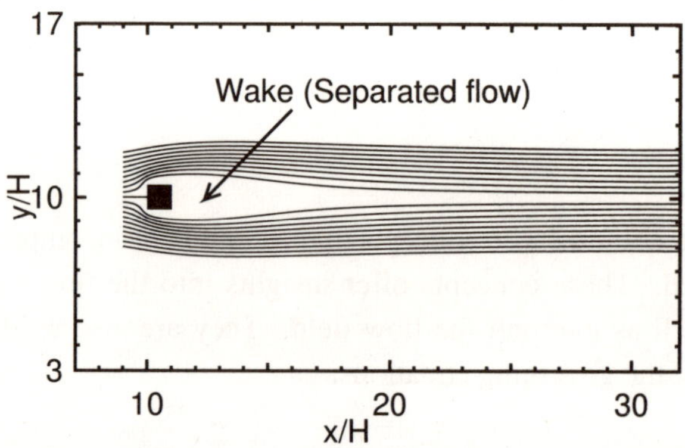

Fig. 2.2: Laminar flow around a square cylinder at $Re = 40$. Flow is from left to right.

In general, flows can be steady or unsteady. In Fig. 2.2, streamlines in the steady flow around a square cylinder placed in a uniform stream are shown. Streamlines in the vicinity of the cylinder alone, not those in the undisturbed freestream, are shown, for the sake of clarity. Also, the cylinder is assumed to be infinitely long in the direction normal to the plane of the paper and so the flow is two-dimensional (*i.e.,* only two velocity components are non-zero). The Reynolds number is defined as $Re = U_\infty H/\nu$, where U_∞ is the freestream velocity (in m/s), H is the height of the cylinder (in m) and ν is the kinematic viscosity of the fluid (in m^2/s). The Reynolds number allows the physical quantities in the flow, namely, the freestream velocity, height of the cylinder and the viscosity of the fluid, to be grouped into a single parameter. This is convenient since different flow fields can be realized by simply changing the Reynolds number without knowing the values of the physical quantities themselves.

It can be surmised from Fig. 2.2 that the flow far away from the cylinder is unaware of the presence of the cylinder, while the flow that approaches the cylinder divides into two and flows over the upper and lower surface. These two streams rejoin some distance downstream of the right face of the square leaving behind a wake region immediately downstream of it. The asymmetry of the flow field about the vertical centerline of the square arises due to the fact that the fluid has a non-zero value for the viscosity. If the viscosity had been zero (hypothetically), then the flow would adhere to the surface of the cylinder. Such a fluid is called an inviscid fluid and flows involving an inviscid fluid are discussed later in Chapter 4. The wake region is separated from the main flow by two streamlines, each one originating from the top right and bottom right corner of the square (not shown in Fig. 2.2). Flow separation and the details of separated flow are discussed later in Chapter 6. It is sufficient to know that the flow itself and the wake in particular, are steady for $Re < 50$ or so.

In general, a flow can be unsteady due to one of the following reasons:

- Imposed unsteady forcing
- Instability
- Transition to turbulence

The pulsatile flow of blood in arteries and veins due to the rhythmic pumping action of the heart, the unsteady flow around the flapping wings of an insect and the flow generated by the undulating body and fins of most marine animals when in motion are examples of flows in the first category. As mentioned earlier, turbulent flows are, unsteady by nature even in the absence of any external forcing mechanism. The turbulent flow issuing from a common house tap is an excellent example. Unsteady flows in the second category are quite interesting, since they are laminar but still exhibit unsteadiness in the absence any imposed unsteady forcing. The flow illustrated in Fig. 2.2, when it becomes unstable, is a good example. This is examined next.

With reference to Fig. 2.2, as the freestream velocity is increased, keeping everything else the same, the separated flow region behind the cylinder becomes longer and more slender. Consequently, it is highly unstable and even an infinitesimal disturbance† can trigger an instability for $Re > 50$.

†In reality, there is always some disturbance present. In the case of experiments, this can

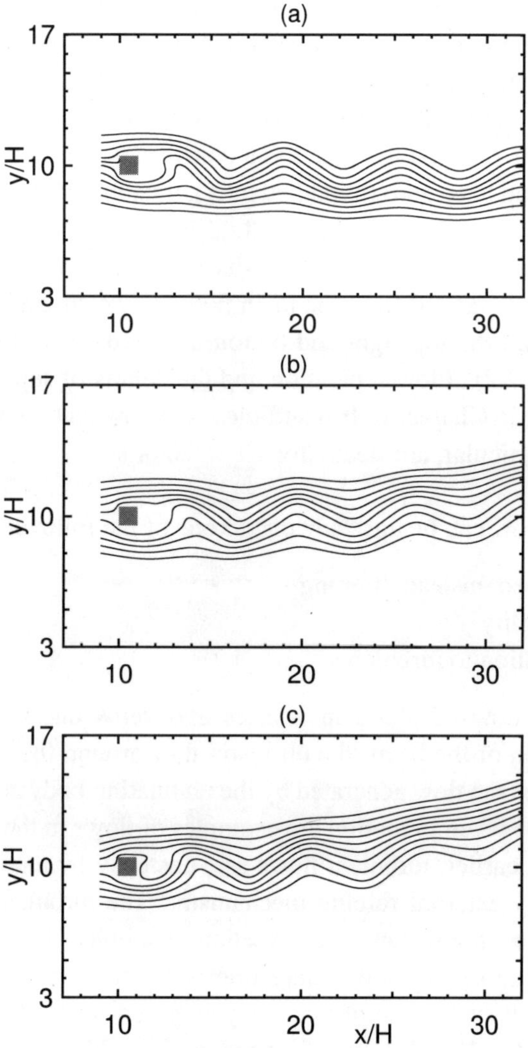

Fig. 2.3: Laminar flow around a square cylinder at $Re = 100$

Once this happens, the separated flow region flaps up and down and is

be a small fluctuation in the freestream velocity or a slight vibration of the setup. In the case of numerical simulations, different errors, namely, the truncation error, discretization error and the round-off error act as the source of disturbance. If these disturbances are completely eliminated, then, in principle, it is possible to delay the onset of the instability. Such a flow is said to be super-critically stable.

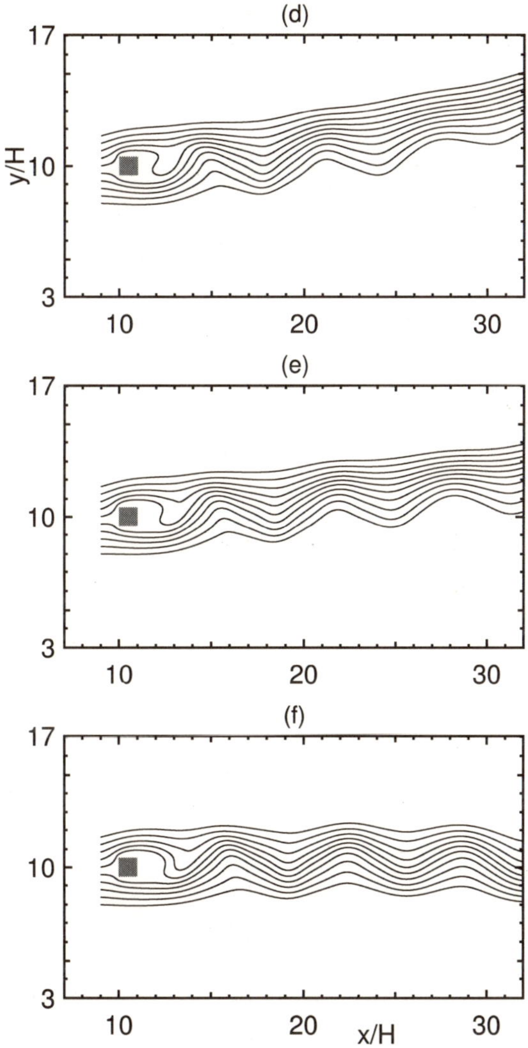

Fig. 2.3: *(cont'd)*Laminar flow around a square cylinder at $Re = 100$

pinched off and shed alternately from the top right and bottom right corners in the form of vortices. These vortices travel downstream and the resulting flow pattern is called the von Kármán vortex street (see cover illustration). It should be noted that the symmetry of the flow about the horizontal centerline of the square is also destroyed at the onset of instability. The

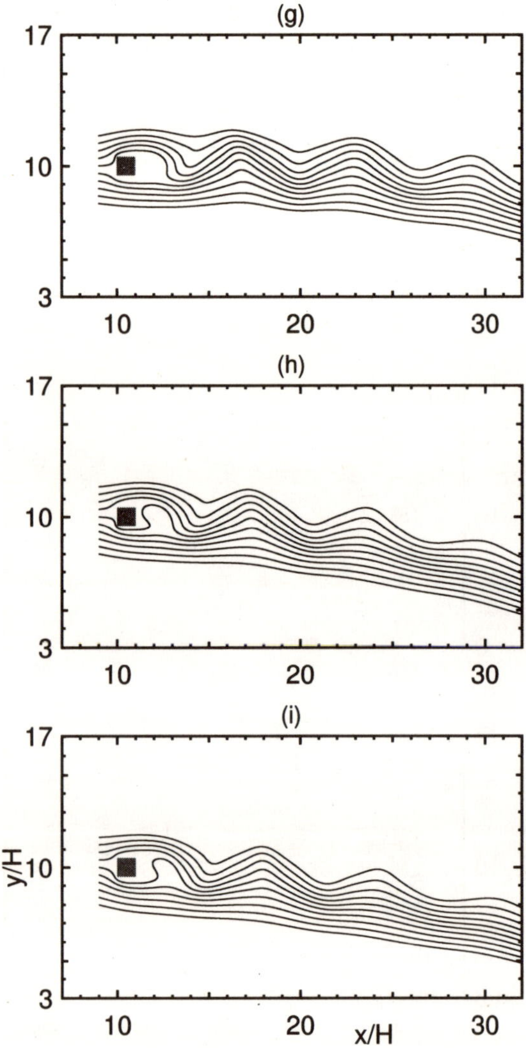

Fig. 2.3: *(cont'd)*Laminar flow around a square cylinder at $Re = 100$

flow field at different instants during a complete vortex shedding cycle is shown in Figs. 2.3(a)-(j). The corresponding Reynolds number is 100. The flapping motion of the wake and the flow further downstream in the vertical direction can be clearly discerned from this sequence of

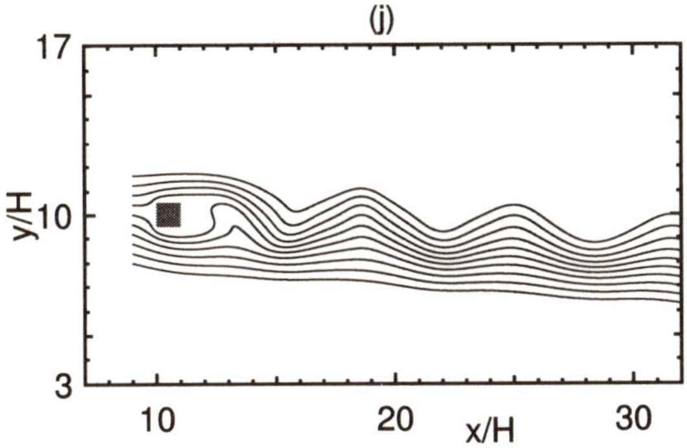

Fig. 2.3: *(cont'd)* Laminar flow around a square cylinder at $Re = 100$

instantaneous streamlines†. When the Reynolds number exceeds 400, the flow becomes three dimensional, *i.e.,* in addition to the unsteady motion in the plane of the paper, there is now an unsteady motion in a direction normal to it. Further still, when the Reynolds number exceeds 1000 or so, the flow transitions from laminar to turbulent. The well defined vortical structures seen in the flow field so far are usually not seen once the flow becomes turbulent.

It is customary, when dealing with unsteady flows, to determine the mean flow, which is the time average of the instantaneous flow fields over a period of time (for example, the instantaneous flow fields in Figs. 2.3(a)-(j)). The time averaged mean flow for the unsteady flow around a square cylinder at $Re = 100$ is shown in Fig. 2.4. It is clear that the averaging process has removed the periodic up and down motion of the flow in he wake and has thus restored the symmetry about the horizontal centerline of the square. The mean flow can be thought of as a "steady flow" since it contains only those features of the flow that persist for long periods of time‡. It follows then that if the mean flow were subtracted from the instantaneous flow, the resulting flow field will reveal regions where the flow is unsteady§. The usefulness of calculating the mean flow arises from

†It should be noted that the cover illustration shows contours of velocity magnitude - not streamlines.

‡Provided the time averaging is done over a sufficiently long period of time.

§The flow illustrated in Fig. 2.3 exhibits unsteadiness only downstream of the square

these two reasons. The contrast between the instantaneous flow field (Fig. 2.3 (a)-(j)) and the mean flow (Fig. 2.4) demonstrates the difficulty in interpreting unsteady flow fields in general. This difficulty becomes more acute when dealing with turbulent flows.

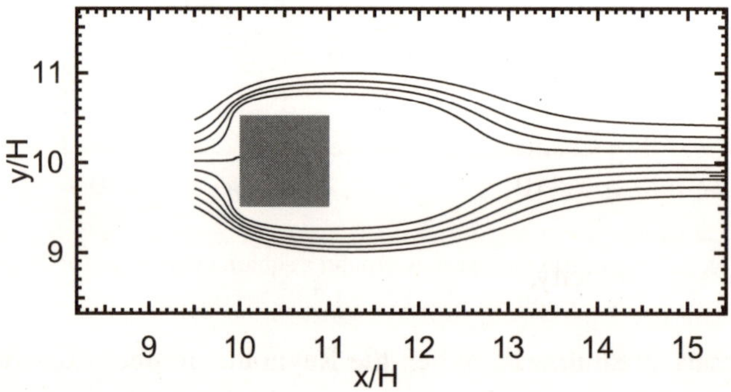

Fig. 2.4: Mean (time averaged) flow around a square cylinder at $Re = 100$. Flow is from left to right.

2.3 Streamlines and pathlines

In our discussion so far, we have utilized streamlines quite extensively without defining it, since the intended meaning is quite intuitive. We now turn to a formal definition of this concept and the manner in which they can be determined for a given velocity field.

A streamline is a continuous line within the flow field such that the tangent vector at any point is parallel to the velocity vector at that point. Mathematically,

$$\frac{dx}{u} = \frac{dy}{v} = \frac{dz}{w} = ds\,, \tag{2.1}$$

where s is the displacement along the streamline. Equation 2.1 describes the path followed by a massless, neutrally buoyant marker particle introduced into the flow field. Since the particle is massless, it attains the fluid

cylinder.

velocity at any location instantly. Starting from a given location x_0, y_0 and z_0, with $s = 0$, Eqn. 2.1 can be integrated to construct the streamline that passes through this location. By repeating this procedure with different starting locations, streamlines covering the entire flow field can be drawn. The streamlines shown in Figs. 2.2, 2.3 and 2.4 have all been drawn in this manner. The following important points regarding streamlines should be noted:

- Streamlines cannot terminate inside a flow field except at an outlet or a stagnation point (where the flow is brought to a complete rest)
- Streamlines cannot intersect except at stagnation point or a point of zero velocity, as otherwise, the tangent to the streamline at the point of intersection will not be unique.
- There is no flow across streamlines.

A careful examination of Eqn. 2.1 will show that, while the velocity components u, v and w can be functions of x, y, z and t, the streamline coordinate s can be a function of x, y and z only. Hence, streamlines can meaningfully depict the flow field at a given instant in time only. Consequently, the above definition has to be modified slightly as follows in order to accommodate unsteady flows.

$$\frac{dx}{u} = \frac{dy}{v} = \frac{dz}{w} = dt\,. \tag{2.2}$$

Starting from a given location x_0, y_0 and z_0 at time $t = 0$, Eqn. 2.2 can be integrated to construct the so-called pathline that passes through this location. As before, by repeating this procedure with different initial locations, numerous pathlines can be constructed. The pathline of a particle released from $x/H = 8.5$ and $y/H = 11.5$ into the unsteady flow discussed so far is shown in Fig. 2.5. Each segment of this pathline is constructed using the appropriate instantaneous flow field shown in Fig. 2.3. A comparison of these two figures clearly brings out the difference between the instantaneous streamline and the pathline. It is, of course obvious, that there is no difference between a streamline and a pathline in the case of a steady flow.

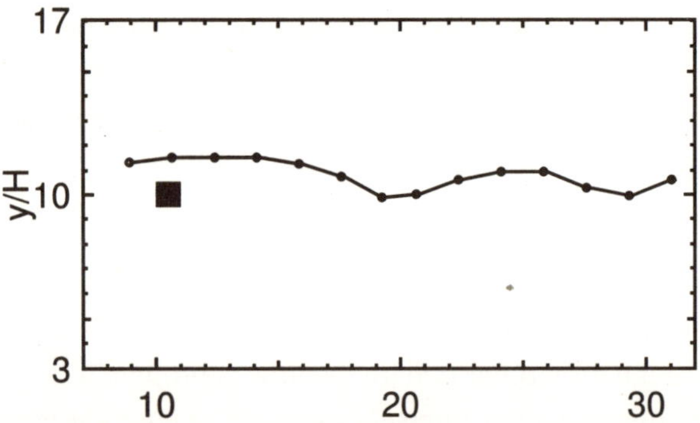

Fig. 2.5: Pathline of a particle released from $x/H = 8.5$ and $y/H = 11.5$ into the flow around a square cylinder at $Re = 100$. Flow is from left to right.

2.4 Stream function, vorticity and circulation

Given a velocity field $\vec{V}(x, y, z, t) = u\hat{i} + v\hat{j} + w\hat{k}$, Eqn. 2.1 allows individual streamlines to be determined, one at a time, for two-dimensional as well as three-dimensional flows. However, an alternative (and insightful) mathematical formulation of the streamline concept is also possible in two-dimensional flows. We start by stipulating that the streamlines on the whole are described by a well defined function $\psi(x, y)$, called the stream function. A careful examination of the second point regarding streamlines in the list given above, suggests that the value of the function $\psi(x, y)$ has to remain constant along a streamline, in order to satisfy this condition. In other words, contours of the function $\psi(x, y)$ are streamlines. The relationship between $\psi(x, y)$ and the velocity field can be determined as follows. From any point (x, y), the change in the value of the stream function due to an arbitrary displacement dx, dy is given by

$$d\psi = \frac{\partial \psi}{\partial x}\, dx + \frac{\partial \psi}{\partial y}\, dy\,. \tag{2.3}$$

If dx, dy are such that the displacement ds is along the streamline passing through (x, y), then

$$d\psi = \frac{\partial \psi}{\partial x} u\; ds + \frac{\partial \psi}{\partial y} v\; ds = \left(\frac{\partial \psi}{\partial x} u + \frac{\partial \psi}{\partial y} v \right) ds\,,$$

after using Eqn. 2.1. In this case, $d\psi = 0$, since the displacement is along a streamline, along which ψ is a constant, as mentioned above. It then follows that,

$$\frac{\partial \psi}{\partial x} u + \frac{\partial \psi}{\partial y} v = 0 \,.$$

Or

$$u = \frac{\partial \psi}{\partial y} \qquad v = -\frac{\partial \psi}{\partial x} \,, \tag{2.4}$$

which are the relations connecting the stream function and the velocity components in Cartesian coordinates. These relations can be written in cylindrical polar coordinates as follows:

$$u_r = \frac{1}{r}\frac{\partial \psi}{\partial \theta} \qquad u_\theta = -\frac{\partial \psi}{\partial r} \,, \tag{2.5}$$

where u_r and u_θ are the radial and the azimuthal components of velocity. Contours of stream function (or streamlines) for three simple flows are shown in Fig. 2.6. The first one is a purely circular motion in which $u_r = 0$ and $u_\theta = r\Omega$, usually called solid body rotation. Here, r is the radius and Ω is the angular velocity. A good real life example is the flow induced in a spinning cup containing a liquid, such as, coffee or tea. If we apply Eqn. 2.5 to this flow, we get

$$\frac{1}{r}\frac{\partial \psi}{\partial \theta} = 0 \qquad -\frac{\partial \psi}{\partial r} = r\Omega \,.$$

It follows from the first equality that ψ is independent of θ and hence the contours of ψ are circles. Upon integrating the second equality, we get $\psi = -\Omega r^2$ after setting $\psi = 0$ at the origin.

The second example in Fig. 2.6, namely, the flow due to a vortex is also a purely circular motion and exhibits circular stream lines. However, in this case, $u_\theta = C/r$, where C is a constant usually called the strength of the

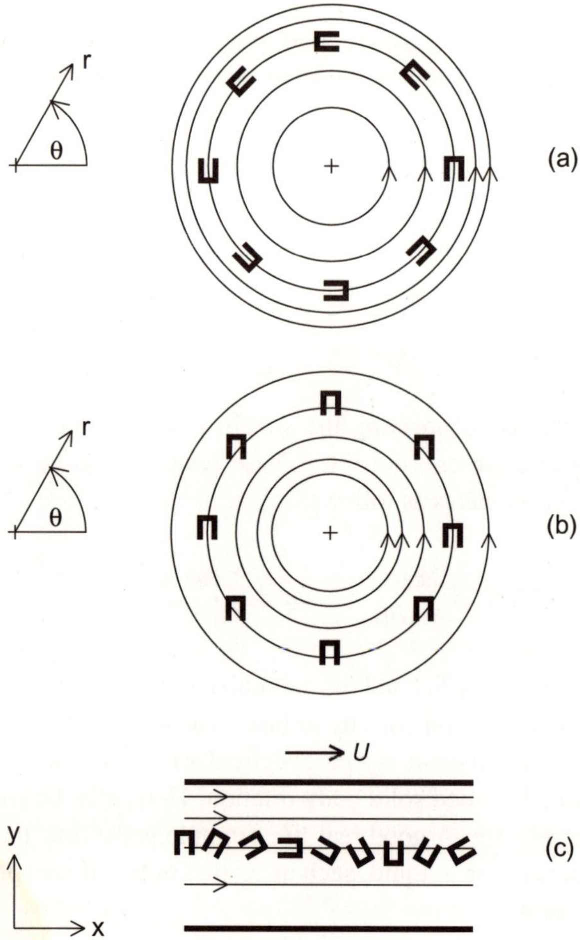

Fig. 2.6: Illustration of vorticity and contours of stream function in different flows. (a) Solid body rotation (b) Vortex flow and (c) Couette flow

vortex. Tornadoes and the flow in a kitchen sink or bathtub as it drains are examples of a vortex flow. Although the $1/r$ variation seems to suggest that the velocity may approach infinity at the center, in reality, in the core region of the vortex, the flow behaves as in a solid body rotation, and so the velocity remains finite. This type of vortex is also called a free vortex. Although the flow fields for these two cases bear a close resemblance to each other, there are differences as well. In the case of the former, the rotational motion of the container induces the fluid flow. In contrast, in the second case, the motion is self-induced. If we follow the same procedure as in the previous case for deriving an equation for ψ, it is easy to show that $\psi = -C\ln(r/r_0)$. Here, we have arbitrarily set $\psi = 0$ at $r = r_0$.

The third example shown in Fig. 2.6 is the so-called Couette flow between two parallel plates. Here, the bottom plate is stationary while the upper one moves to the right with a constant velocity U. The velocity of the flow in this case is given as $u = Uy/H$, where H is the spacing between the plates. Hence the streamlines are horizontal. If we apply Eqn. 2.4 to this flow, we get $\psi = Uy^2/H$ with $\psi = 0$ at $y = 0$.

The streamlines shown in Figs. 2.2, 2.3 and 2.4 have been drawn from a set of starting points located uniformly along a vertical line (rake). At first sight, it appears that the streamlines depict the direction of the velocity vector alone (Eqn. 2.1 also seems to suggest the same thing). However, more information is actually available on account of the fact that there is no flow across a streamline (third point in the list in the previous section). To demonstrate this, we start by combining Eqns. 2.3 and 2.4, with the result

$$d\psi = u\,dy - v\,dx\,. \tag{2.6}$$

Upon integrating this equation between two streamlines as shown in Fig. 2.7, we get

$$\psi_2 - \psi_1 = \int_1^2 u\,dy - v\,dx\,.$$

The path of integration must be known in order to evaluate the right hand side. Without any loss of generality, we consider the four choices shown in Fig. 2.7.

If we evaluate the integral along path B, we get

$$\psi_2 - \psi_1 = \int_1^2 u\,dy\,.$$

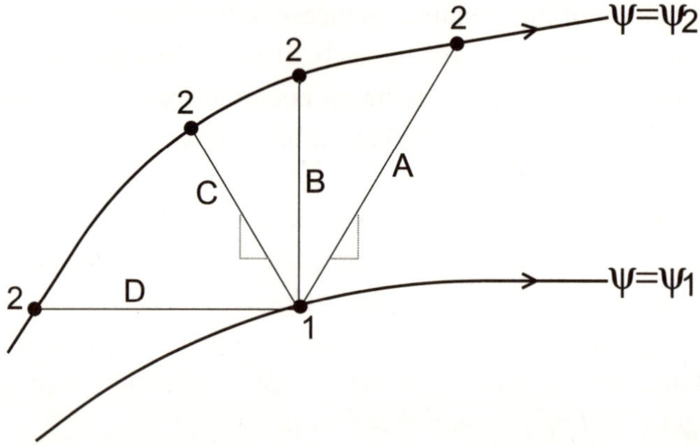

Fig. 2.7: Derivation of the relationship between $\psi_1 - \psi_2$ and the flow rate between the streamlines

On the other hand, if the integration is carried out along path D, then

$$\psi_2 - \psi_1 = -\int_1^2 v \, dx \, .$$

The right hand side in both these expressions is the flow rate between the streamlines. If we imagine paths A and C to be made up of an infinite number of vertical and horizontal segments as shown in Fig. 2.7, then integration along paths A and C essentially amounts to a combination of the integration along paths B and D. Although paths A and C are illustrated as straight lines in Fig. 2.7, the same argument applies even for a curved path. It is thus clear that $\psi_1 - \psi_2$ is equal to the flow rate between the streamlines. As there is no flow across the streamlines, this flow rate must remain the same along the entire length of the streamlines. Hence, streamlines come close when the fluid accelerates and move apart when the fluid decelerates.

Equation 2.4 is not in a convenient form for determining the stream function from a known velocity field, except in simple cases such as the ones illustrated in Fig. 2.6. This can be accomplished for the general case, by relating the stream function to a new quantity called the vorticity. We turn to this next.

Vorticity at a point in the flow field is a vector quantity defined mathematically as

$$\vec{\omega} = \vec{\nabla} \times \vec{u}\,. \tag{2.7}$$

Like the velocity vector, the vorticity vector also has three components, ω_x, ω_y and ω_z in Cartesian coordinates and ω_r, ω_θ and ω_z in cylindrical polar coordinates. Each component points along the corresponding coordinate direction. In this book, since we deal mostly with 2D flows with coordinates as indicated in Fig. 2.6, ω_z alone is non-zero. It is easy show that

$$\omega_z = \frac{\partial v}{\partial x} - \frac{\partial u}{\partial y}\,, \tag{2.8}$$

in Cartesian coordinates, and

$$\omega_z = \frac{1}{r}\frac{\partial\,(r u_\theta)}{\partial r} - \frac{1}{r}\frac{\partial u_r}{\partial \theta}\,, \tag{2.9}$$

in cylindrical polar coordinates. For example, by substituting the expressions for the velocity components given earlier, ω_z for the three flows shown in Fig. 2.6 can be evaluated as 2Ω, 0 and $-U/H$ respectively. It should be noted that the vorticity has the same value everywhere in the flow field for these three cases except that the vorticity is infinite at $r = 0$ for the free vortex case.

If we substitute for u and v in Eqn. 2.8 from Eqn. 2.4, we are led to

$$\frac{\partial^2 \psi}{\partial x^2} + \frac{\partial^2 \psi}{\partial y^2} = -\omega_z\,. \tag{2.10}$$

Similarly, in cylindrical polar coordinates,

$$\frac{\partial^2 \psi}{\partial r^2} + \frac{1}{r}\frac{\partial \psi}{\partial r} + \frac{1}{r^2}\frac{\partial^2 \psi}{\partial \theta^2} = -\omega_z\,. \tag{2.11}$$

Equations 2.10 and 2.11 are Poisson equations for ψ in Cartesian and cylindrical polar coordinates respectively and can be solved using a variety of methods. Hence, given a velocity field (and thus ω_z), the preferred method for obtaining the stream function in the general case is by solving these equations.

The usefulness of the concept of vorticity is not limited to its role in determining the stream function alone. It will be shown mathematically in the next chapter that the vorticity vector at a location represents the rotation of a fluid element at that location. More precisely, each component of the vorticity vector is twice the angular velocity of the fluid element about the coordinate direction corresponding to that component.

The connection between the vorticity and the rotation of a fluid element is illustrated in Fig. 2.6. The vorticity for each one of the flow in Fig. 2.6 has already been calculated to be equal to 2Ω, 0 and $-U/H$ respectively. Consider the inverted U-shaped fluid element in Fig. 2.6(a), which initially ($\theta = 0$) points upwards. As this particle moves along the circular streamline, it is also seen to rotate about its own axis in the anti-clockwise direction (reminiscent of the earth rotating about its own axis while revolving around the sun). In contrast, in Fig. 2.6(b), the particle that starts with the same orientation from the same location, moves along the circular streamline, but without rotating about its own axis. Since the vorticity is non-zero in the first case and zero in the second case, the connection between the rotation of the fluid element about its own axis and vorticity, is clear. In Fig. 2.6(c), the particle which is initially oriented vertically upwards, rotates about is own axis in the clockwise direction as it moves along the horizontal streamline. The rotation of the fluid element demonstrates, once again, that the vorticity is non-zero. We have followed the sign convention that a positive value for the vorticity will lead to a rotation of the fluid element about its own axis in the anti-clockwise direction and *vice versa*. The three flows shown in Fig. 2.6 also illustrate two important facts, namely, circular streamlines do not necessarily imply that vorticity is non-zero and that vorticity can be non-zero even if the streamlines are straight lines. Hence, rotation of a fluid element about its own axis alone (and not about some other axis) can give rise to a non-zero value for the vorticity. An examination of Eqns. 2.8 and 2.9 shows that the rotation or vorticity is generated owing to the gradient in a velocity component in the directions normal to the direction of the velocity component itself.

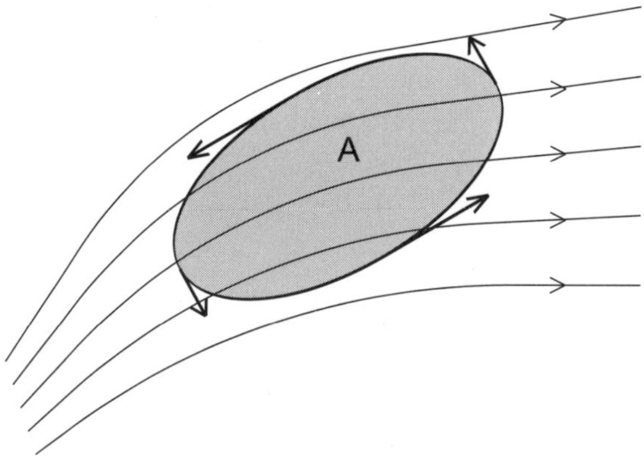

Fig. 2.8: Illustration for Eqn. 2.12. $\hat{t}$ is shown at a few points along the boundary of area A. The outward normal to the area A is perpendicular to the page.

The connection between the vorticity and the orientation of the streamlines can be demonstrated nicely by using the concept of circulation. If we apply Stokes' theorem to Eqn. 2.7, we get †

$$\hat{n} \cdot \vec{\omega} = \lim_{A \to 0} \frac{1}{A} \oint \vec{u} \cdot \hat{t} \, ds \,, \tag{2.12}$$

where, $\hat{n}$ is the outward normal to the area A, $\hat{t}$ is the tangent vector along the boundary of area A and s is the displacement along the edge of A (Fig. 2.8). The cyclic integral is taken along the boundary of this area. The area is shown to be enclosed by an ellipse in Fig. 2.8 for the sake of simplicity alone. In reality, the shape is arbitrary. It is clear from Fig. 2.8 that $\hat{t}$ rotates through 360° for a complete traverse along the edge of A. Hence

† See the book *Div, Grad, Curl and all That* by H. M. Schey (Norton & Co, New York, 1973) for a very nice exposition of this topic.

the dot product in the right hand side of Eqn. 2.12 will change sign once for every complete traverse, in the absence of any change in the direction of $\vec{u}$. In addition, if there is no change in the magnitude of $\vec{u}$ along the path of integration, then the cyclic integral itself will become zero, since the positive and negative contributions will cancel each other exactly[‡]. The cyclic integral will be zero unless $\vec{u}$ changes in direction or magnitude or both during the traverse along the edge of A. As the area A is allowed to shrink to a point, it is easy to extend this reasoning and conclude that a non-zero value for the cyclic integral implies non-zero vorticity at this point. The cyclic integral in the right hand side of Eqn. 2.12 is called the circulation Γ within the area A. Thus,

$$\Gamma = \oint \vec{u} \cdot \hat{t}\, ds\,, \tag{2.13}$$

Evaluation of the circulation Γ for the simple flows illustrated in Fig. 2.6 is demonstrated next. The path of integration to be used for evaluating the integral in Eqn. 2.13 in each case is shown in Fig. 2.9. Since there is complete freedom in the choice of the path, it must be chosen judiciously so that the integration is easy. In general, this means that the individual segments of the path should be along the coordinate directions as shown in Fig. 2.9. For case (a),

$$\Gamma = \int_1^2 \vec{u} \cdot \hat{t}\, ds + \int_2^3 \vec{u} \cdot \hat{t}\, ds + \int_3^4 \vec{u} \cdot \hat{t}\, ds + \int_4^1 \vec{u} \cdot \hat{t}\, ds\,.$$

In cylindrical polar coordinates, $\vec{u} = u_r\hat{e}_r + u_\theta\hat{e}_\theta$ in general, and $\vec{u} = r\Omega\hat{e}_\theta$ in particular for case (a). Also, $\hat{t} = \pm\hat{e}_r$ for segments 1-2 and 3-4 respectively and $\hat{t} = \pm\hat{e}_\theta$ for segments 2-3 and 4-1 respectively. Here, $\hat{e}_r$ and $\hat{e}_\theta$ are the unit vectors in the r and θ directions respectively. Furthermore, $ds = dr$ for segments 1-2, 3-4 and $ds = rd\theta$ for segments 2-3, 4-1. The dot product $\vec{u} \cdot \hat{t}$ is zero along segments 1-2 and 3-4. Hence, the right hand side of the above expression simplifies to

$$\Gamma = r_2^2\Omega \mid \theta_3 - \theta_2 \mid - r_1^2\Omega \mid \theta_1 - \theta_4 \mid = (r_2^2 - r_1^2)\Omega \mid \theta_3 - \theta_2 \mid\,.$$

[‡]This is true only in the case of cartesian coordinates. In the case of polar coordinates, a $1/r$ variation of the tangential velocity will lead to the same result.

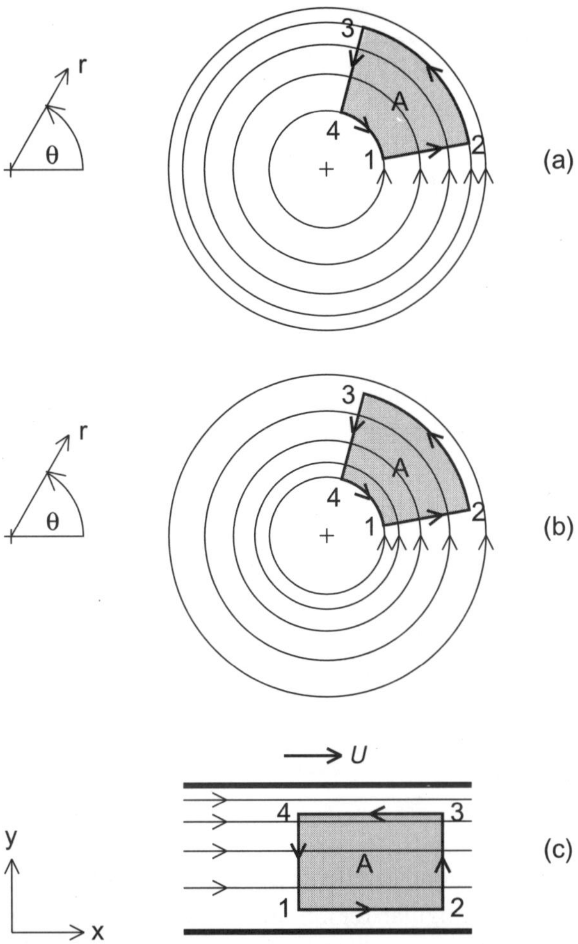

Fig. 2.9: Evaluation of the circulation $\Gamma = \oint \vec{u} \cdot \hat{t}\, ds$ for a few simple flows. $\hat{t}$ is shown at a few points along the edge of area A.

Thus, the circulation within the area A shown in Fig. 2.9(a) is non-zero. The area of the shaded region A in Fig. 2.9(a) can be easily calculated to be $\pi(r_2^2 - r_1^2) \mid \theta_3 - \theta_2 \mid /(2\pi) = (r_2^2 - r_1^2) \mid \theta_3 - \theta_2 \mid /2$. Upon substituting this and the value for Γ calculated above, into Eqn. 2.12, and letting the area shrink to a point, we get $\omega_z = 2\Omega$ at the center of the shaded area A.

Alternatively, if we choose one of the streamlines, say the one passing through point 1, as the path of integration, then

$$\Gamma = \int_0^{2\pi} r_1 \Omega r_1 \, d\theta = 2\pi r_1^2 \Omega \, .$$

The area A is now equal to πr_1^2 and it is easy to show that ω_z at the center of the circular area A is equal to 2Ω. This demonstrates that ω_z has the same value everywhere in the flow field in this case. This, of course, is the same as the value calculated using Eqn. 2.9, as it should be.

Proceeding similarly for case (b), it can be shown that, Γ for the shaded area A and hence the vorticity at its center are both zero. On the other hand, if we choose the streamline passing through 1 as the path of integration, then the circulation $\Gamma = 2\pi C$. Since the area is now πr_1^2, ω_z is equal to $2C/r_1^2$. If we let the area shrink to a point by allowing $r_1 \to 0$, it can be seen that the vorticity at $r = 0$ is infinite. Hence, in case (b), the circulation is zero along any path that does not enclose the origin, and non-zero otherwise. The vorticity is zero everywhere except at the origin, where it is infinite.

For case (c), $\vec{u} \cdot \hat{t}$ is zero along segments 2-3 and 4-1. If the length and the height of the shaded rectangular region are taken as l and h respectively, then it is easy to see that $\Gamma = -Uhl/H$. The area of the shaded area A is lh and so the vorticity at the center of this area is $-U/H$, which is the same as the value obtained by using Eqn. 2.8. It is also easy to show that the vorticity is the same everywhere in the flow field by following the same procedure at different locations.

In general, if there are closed streamlines in the flow (Figs. 2.8(a), (b) and Fig. 2.10), then these regions have non-zero circulation and hence vorticity. However, the converse need not be true. That is, the absence of closed streamlines does not indicate an absence of vorticity, as, for example in Fig. 2.9(c).

2.5 Eulerian and Lagrangian formulations

The equations that govern fluid flow can be derived using either the Eulerian or the Lagrangian approach. The difference between the two formulations is best demonstrated through the following example.

Let us say that information about the traffic pattern in the streets of a city needs to be collected. This can be accomplished in two ways. First, an observer can travel through all the streets in a vehicle, collecting the relevant information along the way. Second, observers can be stationed at various locations along the streets and they can collect the required information. Lagrangian formulation utilizes the first approach, while the Eulerian formulation uses the second approach. It should be realized that, in order to yield reliable data, both the approaches require a large number of observers.

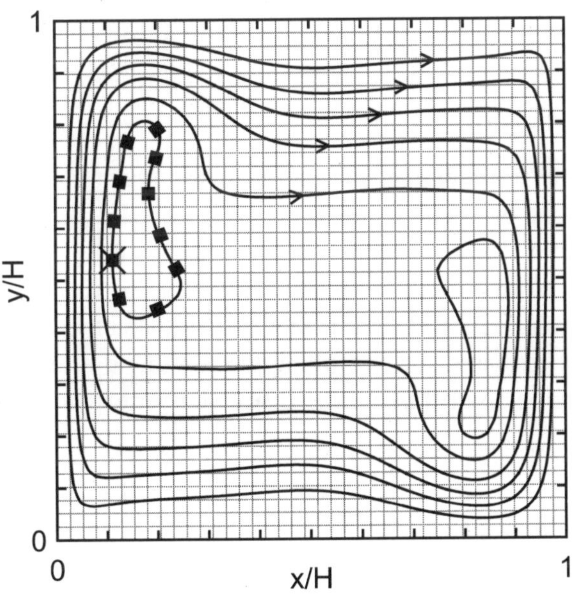

Fig. 2.10: Illustration of Eulerian and Lagrangian framework. The streamlines are due to laminar natural convection in a differentially heated square cavity.

The corresponding situation in the case of a fluid flow is illustrated in Fig. 2.10. Streamlines of the flow in a square cavity in which the left wall is hotter than the right wall, while the top and bottom walls are insulated are shown in Fig. 2.10. The hotter, lighter fluid near the left wall rises due to buoyancy force, turns right when it meets the top wall, sinks down along the right wall as it loses heat and then turns left, thereby setting up a convection loop. Consider the particle (in the form of a square) shown in the figure. In the Lagrangian formulation, the equations that govern the shape and motion of this particle, starting from, say, the point marked with a $\times$ are developed based on the conservation of mass and momentum respectively. It should be noted that, in the figure, only the successive positions of the particle (and not the deformation) are shown, for the sake of clarity. By connecting these positions, the streamline that passes through the starting location of the particle can be obtained. The entire flow field can be described completely by applying this equation to many such particles distributed in the entire flow domain and then solving the resulting set of coupled equations. In order to get an accurate description, a large number of particles have to be used and the displacement of the particles at each instant must be infinitesimally small.

In the Eulerian formulation, the entire flow domain is sub-divided into a large number of small volumes (usually called a differential control volume) such as the gray squares shown in Fig. 2.10. By applying the conservation of mass and momentum principle to the fluid in each of these small volume, the equations that govern the entire flow field can be obtained. Once again, for describing the flow field accurately, the flow domain has to be divided into a very large number of infinitesimally small volumes.

2.6 Material derivative

Material derivative (also called convective derivative or total derivative) is an important concept in fluid mechanics. Consider the flow field illustrated in Fig. 2.10 again. Let the particle be displaced by an amount dx, dy and dz in a time interval dt, from any one of its locations shown in the figure. The change in the temperature dT of the particle owing to this displacement is given by (it must be kept in mind that T is a function of x, y, z and t)

$$dT = \frac{\partial T}{\partial t}dt + \frac{\partial T}{\partial x}dx + \frac{\partial T}{\partial y}dy + \frac{\partial T}{\partial z}dz\,.$$

Since the particle remains on the same streamline, it follows that the displacements dx, dy and dz are such that, $dx = udt$, $dy = vdt$ and $dz = wdt$. Hence, the above equation can be written as

$$\frac{dT}{dt} = \frac{\partial T}{\partial t} + u\frac{\partial T}{\partial x} + v\frac{\partial T}{\partial y} + w\frac{\partial T}{\partial z}\,.$$

This brings out the fact the change in the temperature (or indeed any other scalar quantity such as ρ, u, v or w) of the fluid particle at a given instant is due to a combination of the change in the temperature field with time and space. Thus, the temperature of the particle can change with time, even if the temperature field itself does not. The operator

$$\frac{D}{Dt} \equiv \frac{\partial}{\partial t} + u\frac{\partial}{\partial x} + v\frac{\partial}{\partial y} + w\frac{\partial}{\partial z}\,, \tag{2.14}$$

is called the convective (or material or total) derivative operator. For instance, Du/Dt is the rate of change of the x component of velocity of the particle, which is the acceleration of the particle in the x-direction.

Exercises

(1) A 2D velocity field is given by $u = -y/b^2,\ v = x/a^2$ and $w = 0$. Show that the streamlines are ellipses.

(2) Consider a velocity field (in cylindrical polar coordinates) given by

$$u_r = 0, \quad u_\theta = \frac{\Gamma_0}{2\pi r}\left(1 - e^{-r^2/4\nu t}\right), \quad u_z = 0\,.$$

Determine the acceleration experienced by a particle located at any point in the flow field at time instant 't'. Sketch the streamlines at a given time instant. Assuming $\Gamma_0 = \pi$ and $\nu = 5$, sketch the pathline of a particle released from $r = 1$ and $\theta = 0$ at $t = 0$.

(3) An unsteady flow field is described by the Cartesian velocity components $u = U,\ v = \kappa t,\ w = 0$, where U and κ are constants. Sketch the streamlines at a given time instant and the pathline of a particle released at $x = x_0$.

(4) Consider the unsteady flow with Cartesian velocity components $u = Ue^{-ky}\cos(ky - \Omega t),\ v = 0,\ w = 0$, where $U,\ k$ and Ω are constants and $y > 0$. Sketch the velocity profile and the streamlines at a given time instant assuming $U = k = \Omega = 1$. Sketch the pathline of a particle released at $y = 0.1$ at $t = 0$.

(5) Consider the velocity field $u = Cx,\ v = -Cy,\ w = 0$ with $C = 0.5$. Obtain the equation for the pathline of a particle that is released at $(1,1)$ at $t = 0$. By integrating this equation determine the location of the particle at $t = 1$. Determine the acceleration of the particle at $t = 1$. Show that this value is the same as that obtained using the the total derivative Du/Dt and Dv/Dt.

Chapter 3

The Incompressible Navier-Stokes Equations

In this chapter, the equations that govern the motion of an incompressible fluid are derived. The equations require the flow field to conserve mass and momentum at each point. In the case of flows with temperature variation, energy conservation is also required.

3.1 Continuity equation - Eulerian formulation

In this section, the mass conservation equation, also called the continuity equation is derived using the Eulerian approach. Derivation using the Lagrangian approach is demonstrated following the discussion of the deformation of a fluid element.

Consider the differential control volume in the shape of a parallelopiped shown in Fig. 3.1 with the center located at (x, y, z). The density and the velocity components in the (x, y, z) directions at the center of the differential control volume are ρ and (u, v, w) respectively†. The exact shape of the control volume is immaterial, although this particular choice simplifies the algebra considerably for the Cartesian coordinate system used here. Mass conservation principle for the differential control volume can be simply written as

†We will follow the customary sign convention that a velocity component is positive if it points in the same direction as that of the corresponding coordinate.

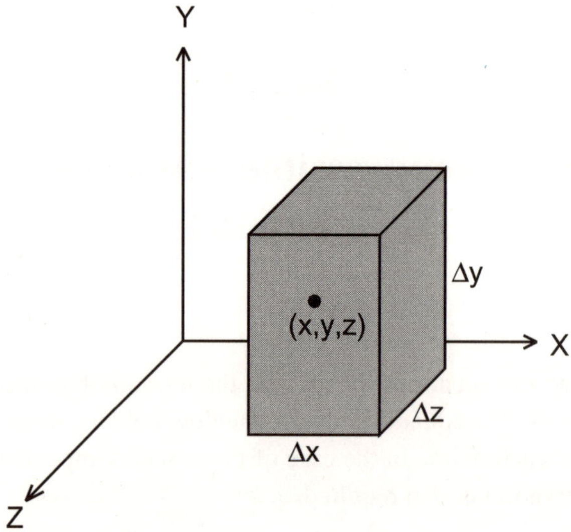

Fig. 3.1: A differential control volume or fluid element with center at (x,y,z)

$$\begin{pmatrix}\text{Rate of change}\\ \text{of mass within}\\ \text{the control}\\ \text{volume}\end{pmatrix} = \begin{pmatrix}\text{Rate at which}\\ \text{mass enters}\\ \text{the control}\\ \text{volume}\end{pmatrix} - \begin{pmatrix}\text{Rate at which}\\ \text{mass leaves}\\ \text{the control}\\ \text{volume}\end{pmatrix}. \tag{3.1}$$

The mass contained within the control volume is equal to $\rho\Delta\mathcal{V}$, where $\Delta\mathcal{V} = \Delta x\,\Delta y\,\Delta z$. The shape and size of the differential control volume and hence $\Delta\mathcal{V}$ does not change in the Eulerian approach. In the general case, when the flow is compressible, we can then write,

$$\begin{pmatrix}\text{Rate of change}\\ \text{of mass within}\\ \text{the control}\\ \text{volume}\end{pmatrix} = \frac{\partial \rho}{\partial t}\Delta x \Delta y \Delta z\,. \tag{3.2}$$

Owing to the sign convention adopted here, it is clear from Fig. 3.1 that fluid enters the differential control volume through the left, bottom and back sides. Similarly, fluid leaves through the right, top and front faces.

Therefore, the first term on the right hand side of Eqn. 3.1 can be written as

$$\begin{pmatrix}\text{Rate at which}\\ \text{mass enters}\\ \text{the control}\\ \text{volume}\end{pmatrix} = \begin{pmatrix}\text{Rate at which}\\ \text{mass enters}\\ \text{through the left}\\ \text{face}\end{pmatrix} + \begin{pmatrix}\text{Rate at which}\\ \text{mass enters}\\ \text{through the bottom}\\ \text{face}\end{pmatrix} + \begin{pmatrix}\text{Rate at which}\\ \text{mass enters}\\ \text{through the back}\\ \text{face}\end{pmatrix},$$

and the second term as

$$\begin{pmatrix}\text{Rate at which}\\ \text{mass leaves}\\ \text{the control}\\ \text{volume}\end{pmatrix} = \begin{pmatrix}\text{Rate at which}\\ \text{mass leaves}\\ \text{through the right}\\ \text{face}\end{pmatrix} + \begin{pmatrix}\text{Rate at which}\\ \text{mass leaves}\\ \text{through the top}\\ \text{face}\end{pmatrix} + \begin{pmatrix}\text{Rate at which}\\ \text{mass leaves}\\ \text{through the front}\\ \text{face}\end{pmatrix}.$$

We now proceed to evaluate the terms on the right hand side of the above two equations, one by one. Since the x-component of velocity at the center of the differential control volume is u, the corresponding value at the left face has to be obtained through a Taylor's series expansion as follows:

$$u(x - \frac{\Delta x}{2}, y, z) = u(x, y, z) - \frac{\Delta x}{2}\frac{\partial u}{\partial x} + O(\Delta x^2)\,,$$

where terms of order Δx^2 and smaller have been neglected, since Δx is itself small[†]. Similarly, the density at the left face is given by

[†]The next term in the series is $\frac{\Delta x^2}{4}\frac{\partial^2 u}{\partial x^2}$. Although Δx^2 is small, for the entire term to be small, $\frac{\partial^2 u}{\partial x^2}$ has to be of order 1. In other words, for the entire term to be small, $\frac{\partial^2 u}{\partial x^2}$ cannot be large.

$$\rho(x - \frac{\Delta x}{2}, y, z) = \rho(x, y, z) - \frac{\Delta x}{2}\frac{\partial \rho}{\partial x} + O(\Delta x^2)\,.$$

Therefore

$$\begin{pmatrix} \text{Rate at which} \\ \text{mass enters} \\ \text{through the left} \\ \text{face} \end{pmatrix} = \left[\rho(x, y, z) - \frac{\Delta x}{2}\frac{\partial \rho}{\partial x} + O(\Delta x^2)\right]$$

$$\times \left[u(x, y, z) - \frac{\Delta x}{2}\frac{\partial u}{\partial x} + O(\Delta x^2)\right] \Delta y \Delta z\,,$$

$$= \left[\rho u - \frac{\Delta x}{2}\rho\frac{\partial u}{\partial x} - \frac{\Delta x}{2}u\frac{\partial \rho}{\partial x} + O(\Delta x^2)\right] \Delta y \Delta z$$

$$= \left[\rho u - \frac{\Delta x}{2}\frac{\partial(\rho u)}{\partial x} + O(\Delta x^2)\right] \Delta y \Delta z\,. \qquad (3.3)$$

The density, y- and the z-component of the velocity on the bottom and the back face can be evaluated similarly. Thus,

$$\begin{pmatrix} \text{Rate at which} \\ \text{mass enters} \\ \text{through the bottom} \\ \text{face} \end{pmatrix} = \left[\rho(x, y, z) - \frac{\Delta y}{2}\frac{\partial \rho}{\partial y} + O(\Delta y^2)\right]$$

$$\times \left[v(x, y, z) - \frac{\Delta y}{2}\frac{\partial v}{\partial y} + O(\Delta y^2)\right] \Delta z \Delta x\,,$$

$$= \left[\rho v - \frac{\Delta y}{2}\frac{\partial(\rho v)}{\partial y} + O(\Delta y^2)\right] \Delta z \Delta x \qquad (3.4)$$

and

$$\begin{pmatrix}\text{Rate at which}\\ \text{mass enters}\\ \text{through the back}\\ \text{face}\end{pmatrix} = \left[\rho(x,y,z) - \frac{\Delta z}{2}\frac{\partial \rho}{\partial z} + O(\Delta z^2)\right]$$

$$\times \left[w(x,y,z) - \frac{\Delta z}{2}\frac{\partial w}{\partial z} + O(\Delta z^2)\right] \Delta x \Delta y\,,$$

$$= \left[\rho w - \frac{\Delta z}{2}\frac{\partial(\rho w)}{\partial z} + O(\Delta z^2)\right] \Delta x \Delta y\,. \tag{3.5}$$

Using similar arguments, it is easy to show that

$$\begin{pmatrix}\text{Rate at which}\\ \text{mass leaves}\\ \text{through the right}\\ \text{face}\end{pmatrix} = \left[\rho u + \frac{\Delta x}{2}\frac{\partial(\rho u)}{\partial x} + O(\Delta x^2)\right] \Delta y \Delta z\,, \tag{3.6}$$

$$\begin{pmatrix}\text{Rate at which}\\ \text{mass leaves}\\ \text{through the top}\\ \text{face}\end{pmatrix} = \left[\rho v + \frac{\Delta y}{2}\frac{\partial(\rho v)}{\partial y} + O(\Delta y^2)\right] \Delta z \Delta x\,, \tag{3.7}$$

and

$$\begin{pmatrix}\text{Rate at which}\\ \text{mass leaves}\\ \text{through the front}\\ \text{face}\end{pmatrix} = \left[\rho w + \frac{\Delta z}{2}\frac{\partial(\rho w)}{\partial z} + O(\Delta z^2)\right] \Delta x \Delta y\,. \tag{3.8}$$

Upon substituting Eqns. 2.2-2.8 into Eqn. 3.1 and dividing throughout by

$\Delta x \Delta y \Delta z$, we get

$$\frac{\partial \rho}{\partial t} + \frac{\partial(\rho u)}{\partial x} + \frac{\partial(\rho v)}{\partial y} + \frac{\partial(\rho w)}{\partial z} = 0\,. \tag{3.9}$$

This is the most general form of the continuity equation. If we expand the product terms in Eqn. 3.9 and collect terms involving ρ, we get

$$\left(\frac{\partial \rho}{\partial t} + u\frac{\partial \rho}{\partial x} + v\frac{\partial \rho}{\partial y} + w\frac{\partial \rho}{\partial z}\right) + \rho\left(\frac{\partial u}{\partial x} + \frac{\partial v}{\partial y} + \frac{\partial w}{\partial z}\right) = 0\,. \tag{3.10}$$

The quantity within the first bracket is convective or material derivative of the density and represents the change in density along a streamline. It is easy to see that if the density remains constant along a streamline, then Eqn. 3.9 becomes

$$\rho\left(\frac{\partial u}{\partial x} + \frac{\partial v}{\partial y} + \frac{\partial w}{\partial z}\right) = 0\,.$$

Since the density takes on non-trivial values, we can finally write,

$$\frac{\partial u}{\partial x} + \frac{\partial v}{\partial y} + \frac{\partial w}{\partial z} = 0 \tag{3.11}$$

which is the continuity equation for an incompressible fluid.

The development above also allows us to write the mathematical condition to be satisfied for a flow to be considered incompressible, namely,

$$\frac{D\rho}{Dt} = \frac{\partial \rho}{\partial t} + u\frac{\partial \rho}{\partial x} + v\frac{\partial \rho}{\partial y} + w\frac{\partial \rho}{\partial z} = 0\,. \tag{3.12}$$

It is important to note that this definition does not require the density to be constant everywhere, but requires it to be a constant only along a streamline. Thus, the density can vary across the streamlines but the flow can still be treated as incompressible. Such a flow is called a stratified flow. However, in most of the incompressible flows, density variations are neglected and hence the density is assumed to be constant in the entire flow field.

Another interesting aspect of the continuity equation, Eqn. 3.11 for incompressible flows is the lack of a time derivative in the equation. Hence, the equation remains the same irrespective of whether the flow is steady or unsteady. This has a profound implication on the mathematical nature of the solutions (or flow fields) of the governing equations. The reader can consult advanced text books on fluid mechanics or partial differential equations for more detailed expositions of this aspect.

3.2 Momentum equation - Lagrangian formulation

In this section, the momentum equation is derived using the Lagrangian approach. Of course, the momentum equation can be derived using the Eulerian approach in the same manner as the continuity equation (the reader is encouraged to try this), but the Lagrangian approach is followed here for the sake of illustration. We start by applying Newton's second law to a particle such as the one shown in Fig. 2.10 (this is shown enlarged in Fig. 3.1) in one of the streamlines. Thus,

$$\begin{pmatrix} \text{Rate of change} \\ \text{of momentum of} \\ \text{the particle} \end{pmatrix} = \begin{pmatrix} \text{Net force} \\ \text{acting on the} \\ \text{particle} \end{pmatrix}. \tag{3.13}$$

Since the mass of the particle $m = \rho \Delta \mathcal{V}$ remains constant[†], the left hand side of the above equation can be written as

$$\begin{pmatrix} \text{Rate of change} \\ \text{of momentum of} \\ \text{the particle} \end{pmatrix} = \begin{pmatrix} \text{Mass of} \\ \text{the} \\ \text{particle} \end{pmatrix} \times \begin{pmatrix} \text{Rate of change} \\ \text{of velocity of} \\ \text{the particle} \end{pmatrix} = \rho \Delta \mathcal{V} \frac{D\vec{u}}{Dt}. \tag{3.14}$$

[†]The density and volume change continuously in such a manner that their product remains constant.

In writing the second equality in Eqn. 3.14, we have used the definition of the material derivative, Eqn. 2.14 and also the fact that Eqn. 3.13 is a vector equation. The left hand side of Eqn. 3.13 is thus completely determined. We turn to the determination of the right hand side next.

3.2.1 *Forces acting on a fluid element and the stress tensor*

Forces acting on a fluid element are of two kinds - body forces and surface forces. Body forces act over the entire volume of the fluid element and do not require any contact. Force due to gravity and buoyancy belong to this category. Without any loss of generality, we neglect the latter and write the net body force acting on the fluid element as

$$\vec{F}_{body} = \rho \Delta \mathcal{V} \vec{g}\,, \tag{3.15}$$

where, $\vec{g}$ is the gravity vector. If the gravity force acts downwards (negative y-direction), then, $\vec{g} = (0, -g, 0)$, with $g = 9.81\, m/s^2$.

Surface forces are the forces exerted on a fluid element by the surrounding fluid. Consequently, these forces act on the surface of the fluid element and arise by virtue of contact. In general, the orientation of the contact area may be such that the outward normal to the contact area does not coincide with the direction of the exerted force. Since two vectors (the outward normal vector and the force vector) are involved, a complete description of the surface force exerted on a fluid element requires the definition of the stress tensor (or the stress matrix). Stress is of course, force divided by the area. The force vector has three components and the area of contact can be projected onto three planes each normal to a coordinate direction. It follows then that the stress tensor has nine components.

The nine components of the stress tensor at a point P are illustrated in Fig. 3.2. These components act along or normal to three mutually perpendicular planes. The components that act along a plane are called shear stresses and are denoted by τ, while those that act normal to a plane are called normal stresses and are denoted by σ. The first subscript denotes the normal

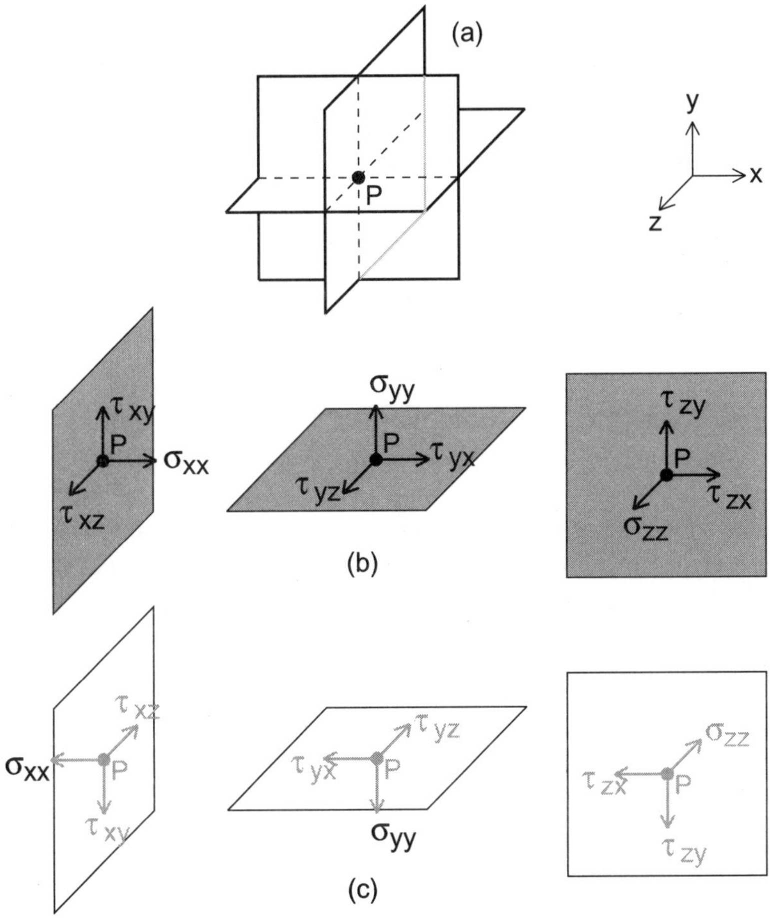

Fig. 3.2: Illustration of the components of the stress tensor at a point P. (a) Three coordinate planes passing through P, (b) components of stress when the normals are in the positive coordinate directions and (c) components of stress when the normals are in the negative coordinate directions

direction of the plane on which the force component acts and the second subscript denotes the direction in which the force component acts†. Hence, the stress tensor at a point can be written as

†The second subscript for the normal stress components is superfluous but it has been retained for the sake of completeness.

$$\bar{\bar{\Pi}} = \begin{pmatrix} \sigma_{xx} & \tau_{xy} & \tau_{xz} \\ \tau_{yx} & \sigma_{yy} & \tau_{yz} \\ \tau_{zx} & \tau_{zy} & \sigma_{zz} \end{pmatrix}, \tag{3.16}$$

where, the double overbars are used to indicate that Π is a tensor. Given the stress tensor $\bar{\bar{\Pi}}$ at a point, the components of stress on a plane in which the point lies and whose normal vector is $\vec{n}$ are given by

$$\vec{\pi} = (\pi_x, \pi_y, \pi_z) = \vec{n} \cdot \bar{\bar{\Pi}} = (n_x \; n_y \; n_z) \begin{pmatrix} \sigma_{xx} & \tau_{xy} & \tau_{xz} \\ \tau_{yx} & \sigma_{yy} & \tau_{yz} \\ \tau_{zx} & \tau_{zy} & \sigma_{zz} \end{pmatrix}.$$

Given the stress tensor at a point, there exist three mutually perpendicular planes, each passing through this point, whose orientations are such that the stress vector on each plane is entirely normal to that plane. Mathematically, this implies that $\vec{\pi} = \sigma\vec{n}$. If we combine this with the expression given above for $\vec{\pi}$, we get

$$\vec{\pi} = \vec{n} \cdot \bar{\bar{\Pi}} = \sigma\vec{n}.$$

This leads to the eigenvalue problem $\left(\bar{\bar{\Pi}} - \sigma\bar{\bar{I}}\right) \cdot \vec{n} = 0$, where $\bar{\bar{I}}$ is the identity matrix (or, the Kronecker delta tensor), σ is the eigenvalue and $\vec{n}$ is the eigenvector. This system of equations admits three pairs of eigenvalues and eigenvectors (mutually perpendicular to each other). The eigenvalues are usually called principal stresses and the eigenvectors are called the principal stress directions.

It is generally very convenient to write the stress tensor $\bar{\bar{\Pi}}$ as the sum of a spherical part and a deviatoric part $\bar{\bar{\Pi}}'$, thus

$$\bar{\bar{\Pi}} = \begin{pmatrix} -p & 0 & 0 \\ 0 & -p & 0 \\ 0 & 0 & -p \end{pmatrix} + \bar{\bar{\Pi}}',$$

and

$$\bar{\bar{\Pi}}' = \begin{pmatrix} \sigma'_{xx} & \tau_{xy} & \tau_{xz} \\ \tau_{yx} & \sigma'_{yy} & \tau_{yz} \\ \tau_{zx} & \tau_{zy} & \sigma'_{zz} \end{pmatrix}, \qquad (3.17)$$

where, $\sigma'_{xx} = \sigma_{xx} + p$, $\sigma'_{yy} = \sigma_{yy} + p$ and $\sigma'_{zz} = \sigma_{zz} + p$. The significance of this decomposition will be demonstrated later . It should be noted that the principal directions of the deviatoric stress tensor are identical to those of the stress tensor itself.

Consider the fluid element shown in Fig. 3.1 again. The components of the stress tensor at the center of this element are given by Eqn. 3.16. The components of stress on the right and left faces of the fluid element can be obtained by using a Taylor's expansion as

$$\sigma_{xx}\big|_{x\pm\frac{\Delta x}{2}} = \sigma_{xx} \pm \frac{\partial \sigma_{xx}}{\partial x}\frac{\Delta x}{2} + O(\Delta x^2),$$

$$\tau_{xy}\big|_{x\pm\frac{\Delta x}{2}} = \tau_{xy} \pm \frac{\partial \tau_{xy}}{\partial x}\frac{\Delta x}{2} + O(\Delta x^2),$$

$$\tau_{xz}\big|_{x\pm\frac{\Delta x}{2}} = \tau_{xz} \pm \frac{\partial \tau_{xz}}{\partial x}\frac{\Delta x}{2} + O(\Delta x^2),$$

where the negative sign is applicable for the left face. The components of stress on the top and bottom face can be written as

$$\sigma_{yy}\big|_{y\pm\frac{\Delta y}{2}} = \sigma_{yy} \pm \frac{\partial \sigma_{yy}}{\partial y}\frac{\Delta y}{2} + O(\Delta y^2)$$

$$\tau_{yx}\big|_{y\pm\frac{\Delta 2}{2}} = \tau_{yx} \pm \frac{\partial \tau_{yx}}{\partial y}\frac{\Delta y}{2} + O(\Delta y^2)$$

$$\tau_{yz}\big|_{y\pm\frac{\Delta y}{2}} = \tau_{yz} \pm \frac{\partial \tau_{yz}}{\partial y}\frac{\Delta y}{2} + O(\Delta y^2),$$

where the negative sign is applicable for the bottom face. The components of stress on the front and back face can be written as

$$\sigma_{zz}\big|_{z\pm\frac{\Delta z}{2}} = \sigma_{zz} \pm \frac{\partial \sigma_{zz}}{\partial z}\frac{\Delta z}{2} + O(\Delta z^2)\,,$$

$$\tau_{zx}\big|_{z\pm\frac{\Delta z}{2}} = \tau_{zx} \pm \frac{\partial \tau_{zx}}{\partial z}\frac{\Delta z}{2} + O(\Delta z^2)\,,$$

$$\tau_{zy}\big|_{z\pm\frac{\Delta z}{2}} = \tau_{zy} \pm \frac{\partial \tau_{zy}}{\partial z}\frac{\Delta z}{2} + O(\Delta z^2)\,,$$

where the negative sign is applicable for the back face. Thus, the net force on the fluid element arising from these stresses in the x, y and z directions are, respectively,

$$\begin{aligned} &\left(\frac{\partial \sigma_{xx}}{\partial x} + \frac{\partial \tau_{yx}}{\partial y} + \frac{\partial \tau_{zx}}{\partial z}\right)\Delta\mathcal{V}\,, \\ &\left(\frac{\partial \sigma_{yy}}{\partial y} + \frac{\partial \tau_{xy}}{\partial x} + \frac{\partial \tau_{zy}}{\partial z}\right)\Delta\mathcal{V}\,, \\ &\left(\frac{\partial \sigma_{zz}}{\partial z} + \frac{\partial \tau_{xz}}{\partial x} + \frac{\partial \tau_{yz}}{\partial y}\right)\Delta\mathcal{V}\,. \end{aligned} \tag{3.18}$$

Upon substituting Eqns. 3.14, 3.15 and 3.18 into Eqn. 3.13 and dividing by $\Delta\mathcal{V}$, we get

$$\begin{aligned} \rho\frac{Du}{Dt} &= -\frac{\partial p}{\partial x} + \frac{\partial \sigma'_{xx}}{\partial x} + \frac{\partial \tau_{yx}}{\partial y} + \frac{\partial \tau_{zx}}{\partial z}\,, \\ \rho\frac{Dv}{Dt} &= -\rho g - \frac{\partial p}{\partial y} + \frac{\partial \sigma'_{yy}}{\partial y} + \frac{\partial \tau_{xy}}{\partial x} + \frac{\partial \tau_{zy}}{\partial z}\,, \\ \rho\frac{Dw}{Dt} &= -\frac{\partial p}{\partial z} + \frac{\partial \tau_{xz}}{\partial x} + \frac{\partial \tau_{yz}}{\partial y} + \frac{\partial \sigma'_{zz}}{\partial z}\,, \end{aligned} \tag{3.19}$$

where, we have used Eqn. 3.17 also. In the context of Eqn. 3.19, p is the pressure of the fluid. It is the same as the thermodynamic pressure used in the equation of state for the fluid. In the absence of any flow, *i.e.,* for a fluid at rest, the left hand side of Eqn. 3.19 is zero, since $\vec{u} = 0$. Also, for such a situation, the fluid pressure balances the hydrostatic force. Hence the components of the deviatoric part of the stress tensor go to zero when the fluid is at rest. Also, the off-diagonal elements of the deviatoric part of the stress tensor (and indeed the stress tensor itself), which are shear stresses, are non-zero only in the presence of a flow. It can then be inferred that any imposed shear stress will induce a flow. It may be recalled that this was presented as the definition of a fluid in Section 2.1.

An important aspect concerning the stress tensor, namely, symmetry, is discussed next. Consider the fluid element shown in Fig. 3.1 and the forces acting on its faces. It can be seen that, for example, there is a torque acting on the fluid element about its own z axis arising from the shear forces on the right, left and top, bottom faces (see Fig. 3.2). It should be clear from the direction of these forces, that the shear forces alone (and not the normal forces), can generate a torque. The couple (or torque) due to the shear forces on the right and left faces is equal to (with the same sign convention as that for the vorticity)

$$\left(\tau_{xy} + \frac{\partial \tau_{xy}}{\partial x}\frac{\Delta x}{2}\right)(\Delta y \Delta z)\frac{\Delta x}{2} + \left(\tau_{xy} - \frac{\partial \tau_{xy}}{\partial x}\frac{\Delta x}{2}\right)(\Delta y \Delta z)\frac{\Delta x}{2} .$$

Upon simplification, this reduces to $\tau_{xy}\Delta\mathcal{V}$. The couple due to the shear forces on the top and bottom faces can be similarly shown to be $\tau_{yx}\Delta\mathcal{V}$. The combined torque induces an angular velocity Ω_z on the fluid element. Thus,

$$\mathcal{I}_z \frac{d\Omega_z}{dt} = (\tau_{xy} - \tau_{yx})\,\Delta\mathcal{V} ,$$

where, $\mathcal{I}_z$ is the moment of inertia of the fluid element about an axis which passes through its center and parallel to the z axis. It is easy to verify from any basic text book in mechanics that $\mathcal{I}_z = (1/12)\rho\Delta\mathcal{V}(\Delta x^2 + \Delta y^2)$. Upon substituting this expression into the equation above and simplifying, we get

$$\frac{1}{12}\rho\frac{d\Omega_z}{dt}\left(\Delta x^2 + \Delta y^2\right) = (\tau_{xy} - \tau_{yx}) .$$

If the fluid element is now allowed to shrink to a point by letting Δx, Δy and $\Delta z \rightarrow 0$, then $d(\Omega_z)/dt$ becomes infinite unless, $\tau_{xy} = \tau_{yx}$. Using similar arguments, it is easy to establish that, $\tau_{yz} = \tau_{zy}$ and $\tau_{zx} = \tau_{xz}$. Since a purely rotational motion of a fluid element about its own axes (vorticity) does not result in any deformation (or, straining), it follows that, shear stresses, which are necessarily accompanied by a straining of the fluid element, cannot give rise to such a motion. The derivation above demonstrates that the symmetry of the tensor arises from this consideration.

The next step in the development of Eqn. 3.19 is to relate the components of the deviatoric stress tensor to the velocity field, through the components of the strain rate tensor†.

3.3 Straining of a fluid element and the strain rate tensor

Consider again the fluid element shown in Fig. 3.1. The motion and deformation of the particle after a time interval Δt is illustrated in Fig. 3.3. The location, shape and orientation of the fluid element at time $t + \Delta t$ is determined by the combined outcome of the following:

- pure translation
- pure rotation
- distortion (shear)
- volume dilatation

Let us now consider each of these in detail. For the sake of simplicity, and without any loss of generality, we will focus our attention on the face marked $ABCD$ in this figure. At a time instant $t + \Delta t$, the position of these vertices are marked as A', B', C' and D' in Fig. 3.3. The changes in the positions of the vertices are the result of (a) motions without any deformation and (b) deformation of the fluid element. These are illustrated in Fig. 3.4.

---

†In solid mechanics, the stress tensor is related to the strain tensor. However, in fluid mechanics, since the fluid begins to flow due to the applied stress, the strain rate tensor, rather than the strain tensor is appropriate.

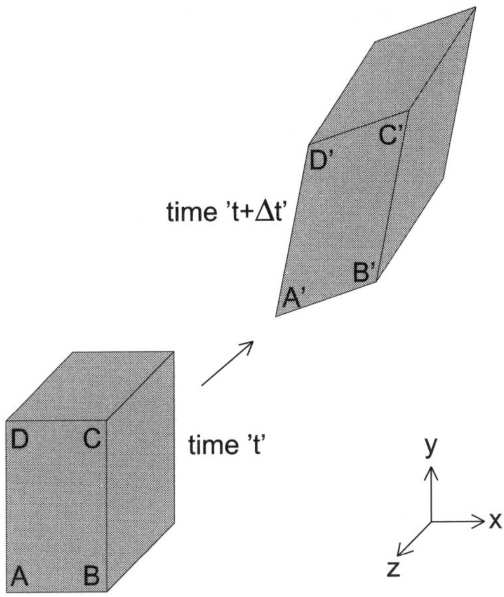

Fig. 3.3: Illustration of the motion and deformation of a fluid element

It can be inferred from Fig. 3.4 that a translation or rotation of the fluid element does not result in a deformation of the fluid element. It should be noted that, in the latter case, the diagonal rotates, while initially perpendicular lines such as AB and AD continue to remain perpendicular, from which the above inference is drawn. In addition, the edges AB and AD rotate at the same rate and in the same direction, which results in an overall rotation of the diagonal. On the other hand, the fluid element deforms under a shear or dilatation. As seen in Fig. 3.4, a pure shearing of the fluid element results in a distortion of the fluid element. It is important to note that this distortion occurs without any net rotation, since the diagonals AC and $A'C'$ coincide. However, the edges $A'D'$ and $A'B'$, which were initially perpendicular, are no longer so. These two edges rotate at the same rate but in opposite directions so that there is no net rotation of the diagonal. Such a shearing strain would arise from the action of the τ_{xy} and τ_{yx} components of the stress tensor (see Fig. 3.2) on the fluid element. Dilatational strain of the fluid element is shown in Fig. 3.4 as an elongation in the x-direction and a compression in the y-direction, which

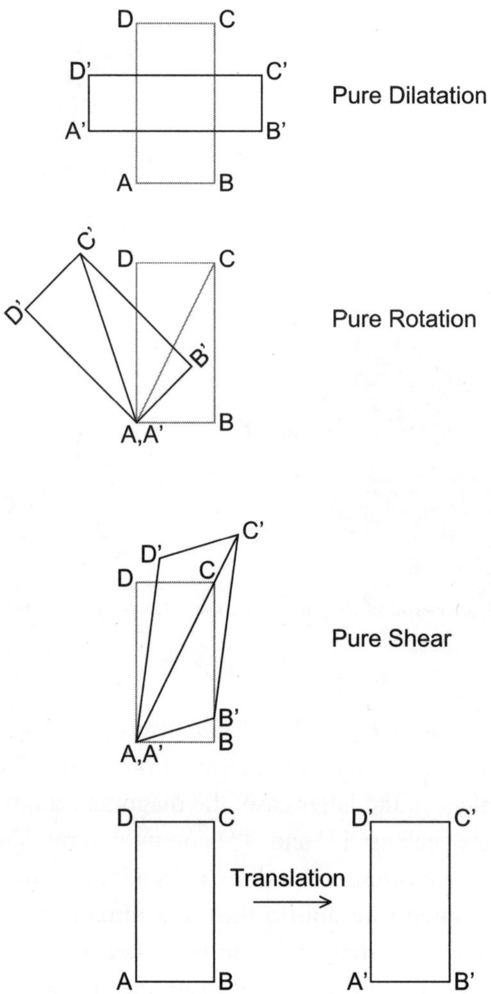

Fig. 3.4: Illustration of the motion and straining of a fluid element

would result in a change in the volume. It is easy to see that such a strain would be the result of the σ_{xx} or σ_{yy} (or both) components of the stress tensor (see Fig. 3.2). In general, the off-diagonal components of the stress tensor cause shear strains (which are the off-diagonal components of the strain rate tensor) and the diagonal components cause dilatational strains (which are the diagonal components of the strain rate tensor). We turn to a quantitative evaluation of these strains next.

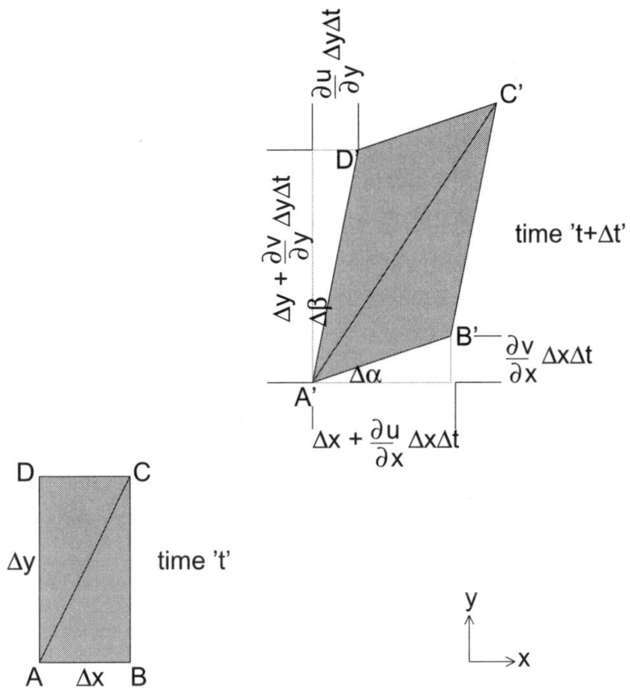

Fig. 3.5: Illustration of the deformation of a fluid element

Consider the face $ABCD$ of the fluid element shown again in Fig. 3.5. At time instant t, let the velocity components at vertex A be u and v along the x and y coordinate directions respectively. The velocity components at the vertices B', C' and D' are then given as

$$\left(u + \frac{\partial u}{\partial x}\Delta x,\, v + \frac{\partial v}{\partial x}\Delta x\right),$$

$$\left(u + \frac{\partial u}{\partial x}\Delta x + \frac{\partial u}{\partial y}\Delta y,\, v + \frac{\partial v}{\partial x}\Delta x + \frac{\partial v}{\partial y}\Delta y\right)$$

and

$$\left(u + \frac{\partial u}{\partial y}\Delta y,\, v + \frac{\partial v}{\partial y}\Delta y\right)$$

respectively. Here, we have ignored terms of order Δx^2 and Δy^2. Vertex A is displaced by an amount $u\Delta t$ and $v\Delta t$ along the x- and y-directions after a time interval of Δt to a location A'. The locations of the other vertices relative to A' are shown in Fig. 3.5.

Rotation of the fluid element is quantified by the angular velocity about each coordinate axis. With reference to Fig. 3.5, the angular velocity of the element (or face) about the z-axis can be evaluated as the arithmetic mean of the rates of rotation of edges AD and AB. Thus,

$$\Omega_z = \frac{1}{2}\left(\frac{d\alpha}{dt} - \frac{d\beta}{dt}\right),$$

where, the sign convention for angular displacement is the same as before (*i.e.*, rotation in the anti-clockwise direction is positive). For small angular displacements, we can write

$$\Delta\alpha = \frac{\frac{\partial v}{\partial x}\Delta x \Delta t}{\Delta x} = \frac{\partial v}{\partial x}\Delta t$$

and

$$\Delta\beta = \frac{\frac{\partial u}{\partial y}\Delta y \Delta t}{\Delta y} = \frac{\partial u}{\partial y}\Delta t\,.$$

Thus,

$$\Omega_z = \frac{1}{2}\left(\frac{\partial v}{\partial x} - \frac{\partial u}{\partial y}\right) = \frac{\omega_z}{2}\,.$$

This expression explicitly shows the relationship between the vorticity and the angular velocity of the fluid element. Similar expressions can be derived for Ω_x and Ω_y.

Shear strain rate of the fluid element is defined as the average rate of decrease of the angle between two initially perpendicular lines (AB and AD). Thus

$$\epsilon_{xy} = \frac{1}{2}\left(\frac{d\alpha}{dt} + \frac{d\beta}{dt}\right).$$

If we substitute for $d\alpha/dt$ and $d\beta/dt$ from above, we get

$$\epsilon_{xy} = \frac{1}{2}\left(\frac{\partial v}{\partial x} + \frac{\partial u}{\partial y}\right).$$

It should be noted that $\epsilon_{xy} = \epsilon_{yx}$. Similarly,

$$\epsilon_{yz} = \frac{1}{2}\left(\frac{\partial w}{\partial y} + \frac{\partial v}{\partial z}\right)$$

and

$$\epsilon_{zx} = \frac{1}{2}\left(\frac{\partial u}{\partial z} + \frac{\partial w}{\partial x}\right).$$

Dilatational strain of a fluid element is the rate of change of length along each coordinate direction divided by the original length. If we consider the edge AB in Fig. 3.5, then the dilatational strain along the x-direction is given as

$$\epsilon_{xx} = \frac{1}{\Delta t}\frac{\Delta x + \frac{\partial u}{\partial x}\Delta x \Delta t - \Delta x}{\Delta x} = \frac{\partial u}{\partial x}.$$

If we consider the edge $A'D'$, then the dilatational strain along the y-direction, ϵ_{yy}, comes out to be equal to $\frac{\partial v}{\partial y}$. It is an easy extension to show that $\epsilon_{zz} = \frac{\partial w}{\partial z}$.

The strain rate tensor can finally be written as

$$\bar{\bar{\epsilon}} = \begin{pmatrix} \epsilon_{xx} & \epsilon_{xy} & \epsilon_{xz} \\ \epsilon_{yx} & \epsilon_{yy} & \epsilon_{yz} \\ \epsilon_{zx} & \epsilon_{zy} & \epsilon_{zz} \end{pmatrix}$$

$$= \begin{pmatrix} \dfrac{\partial u}{\partial x} & \dfrac{1}{2}\left(\dfrac{\partial v}{\partial x} + \dfrac{\partial u}{\partial y}\right) & \dfrac{1}{2}\left(\dfrac{\partial u}{\partial z} + \dfrac{\partial w}{\partial x}\right) \\ \dfrac{1}{2}\left(\dfrac{\partial v}{\partial x} + \dfrac{\partial u}{\partial y}\right) & \dfrac{\partial v}{\partial y} & \dfrac{1}{2}\left(\dfrac{\partial w}{\partial y} + \dfrac{\partial v}{\partial z}\right) \\ \dfrac{1}{2}\left(\dfrac{\partial u}{\partial z} + \dfrac{\partial w}{\partial x}\right) & \dfrac{1}{2}\left(\dfrac{\partial w}{\partial y} + \dfrac{\partial v}{\partial z}\right) & \dfrac{\partial w}{\partial z} \end{pmatrix} . \quad (3.20)$$

We note in passing that the strain rate tensor is also symmetric, as otherwise, it cannot be related to the stress tensor, which is symmetric.

3.4 Relation between the deviatoric stress tensor and the strain rate tensor

The relation between the components of the deviatoric stress tensor and the strain rate tensor as proposed by Stokes obeys the following postulates:

- The components of the deviatoric stress tensor depend linearly on the components of the strain rate tensor, similar to the Hooke's law of elasticity in solid mechanics.
- This relationship does not change with a rotation (or exchange) of coordinates.
- The principal axes of the deviatoric stress tensor and the strain rate tensor are the same.

The last two items in the above list ensure isotropy *i.e.,* the fluid deforms in the same manner in all the coordinate directions. It can be shown that these postulates require†

$$\bar{\bar{\Pi}}' = 2\mu\,\bar{\bar{\epsilon}} + \lambda\,(\vec{\nabla}\cdot\vec{u})\,\bar{\bar{I}}\,, \quad (3.21)$$

where, μ is the familiar dynamic viscosity coefficient and λ is the not so familiar second coefficient of viscosity or the coefficient of bulk viscosity.

†The interested reader is referred to the advanced text books listed at the end

It should be noted that the right hand side of Eqn. 3.21 vanishes when there is no flow, as it should.

3.4.1 *Stokes' hypothesis*

Equation 3.21 can be rewritten as

$$\bar{\bar{\Pi}} = -p\,\bar{\bar{I}} + 2\mu\,\bar{\bar{\epsilon}} + \lambda\,(\vec{\nabla}\cdot\vec{u})\,\bar{\bar{I}}\,.$$

It follows that

$$\sigma_{xx} = -p + 2\mu\frac{\partial u}{\partial x} + \lambda(\vec{\nabla}\cdot\vec{u})\,,$$

$$\sigma_{yy} = -p + 2\mu\frac{\partial v}{\partial y} + \lambda(\vec{\nabla}\cdot\vec{u})$$

and

$$\sigma_{zz} = -p + 2\mu\frac{\partial w}{\partial z} + \lambda(\vec{\nabla}\cdot\vec{u})\,.$$

The mechanical pressure $\bar{p}$ is customarily defined as $-(\sigma_{xx}+\sigma_{yy}+\sigma_{zz})/3$. Upon substituting for the σ's, we get

$$\bar{p} = p + (2\mu + 3\lambda)\vec{\nabla}\cdot\vec{u}\,.$$

It is important to keep in mind that we have not assumed anything about the nature of the fluid (compressible or incompressible) so far in the development related to the stress tensor and the strain rate tensor. We do so now and note that from Eqn. 3.11, for incompressible flows, $\vec{\nabla}\cdot\vec{u} = 0$. It follows then that the mechanical pressure $\bar{p}$ and the thermodynamic pressure are equal. Fortunately, this relieves us from the burden of having to determine λ. However, in the general case, $\vec{\nabla}\cdot\vec{u} \neq 0$ and as pointed out by Stokes, the mechanical pressure and the thermodynamic pressure are not the same. In addition, λ also has to be known. Stokes resolved these two issues by an *ad-hoc* hypothesis that $\lambda = -(2/3)\mu$. Although this hypothesis lacks a theoretical or experimental basis, the resulting equations have been shown to describe fluid flows (both incompressible and compressible) extremely well. It appears now that this success has less to do with the validity of the hypothesis and more to do with the fact that $\vec{\nabla} \cdot \vec{u}$ is quite small in most flows. The reader is referred to the book *Viscous Fluid Flow* by White for an interesting discussion of this point.

3.5 Incompressible Navier-Stokes equations

Upon substituting Eqn. 3.21 into Eqn. 3.19, and after using the fact that, for incompressible flows, $\vec{\nabla} \cdot \vec{u} = 0$, we finally get

$$\begin{aligned}
\rho \frac{Du}{Dt} &= -\frac{\partial p}{\partial x} + \frac{\partial}{\partial x}\left(2\mu \frac{\partial u}{\partial x}\right) + \frac{\partial}{\partial y}\left(\mu\left(\frac{\partial v}{\partial x} + \frac{\partial u}{\partial y}\right)\right) \\
&\quad + \frac{\partial}{\partial z}\left(\mu\left(\frac{\partial w}{\partial x} + \frac{\partial u}{\partial z}\right)\right), \\
\rho \frac{Dv}{Dt} &= -\rho g - \frac{\partial p}{\partial y} + \frac{\partial}{\partial x}\left(\mu\left(\frac{\partial u}{\partial y} + \frac{\partial v}{\partial x}\right)\right) + \frac{\partial}{\partial y}\left(2\mu \frac{\partial v}{\partial y}\right) \\
&\quad + \frac{\partial}{\partial z}\left(\mu\left(\frac{\partial w}{\partial y} + \frac{\partial v}{\partial z}\right)\right), \\
\rho \frac{Dw}{Dt} &= -\frac{\partial p}{\partial z} + \frac{\partial}{\partial x}\left(\mu\left(\frac{\partial u}{\partial z} + \frac{\partial w}{\partial x}\right)\right) + \frac{\partial}{\partial y}\left(\mu\left(\frac{\partial v}{\partial z} + \frac{\partial w}{\partial y}\right)\right) \\
&\quad + \frac{\partial}{\partial z}\left(2\mu \frac{\partial w}{\partial z}\right).
\end{aligned} \tag{3.22}$$

Equation 3.22 can be simplified further if the viscosity remains constant. This leads to

$$\begin{aligned}
\rho \frac{Du}{Dt} &= -\frac{\partial p}{\partial x} + \mu\left(\frac{\partial^2 u}{\partial x^2} + \frac{\partial^2 u}{\partial y^2} + \frac{\partial^2 u}{\partial z^2}\right), \\
\rho \frac{Dv}{Dt} &= -\rho g - \frac{\partial p}{\partial y} + \mu\left(\frac{\partial^2 v}{\partial x^2} + \frac{\partial^2 v}{\partial y^2} + \frac{\partial^2 v}{\partial z^2}\right), \\
\rho \frac{Dw}{Dt} &= -\frac{\partial p}{\partial z} + \mu\left(\frac{\partial^2 w}{\partial x^2} + \frac{\partial^2 w}{\partial y^2} + \frac{\partial^2 w}{\partial z^2}\right),
\end{aligned} \tag{3.23}$$

where, we have again used $\vec{\nabla} \cdot \vec{u} = 0$. Equations 3.23 are called the incompressible Navier-Stokes equations. These equations are given in the Appendix in cylindrical polar coordinates.

3.6 Newtonian and Non-Newtonian fluids

From Eqn. 3.21, we can write

$$\tau_{yx} = 2\mu\epsilon_{yx} \,. \tag{3.24}$$

Any fluid that obeys this relationship is usually called a Newtonian fluid. Most commonly encountered fluids such as air, water, oil *etc.,* are Newtonian fluids. Other fluids such as honey, paint, ketchup, polymers *etc.,* do not obey Eqn. 3.24 and these are called non-Newtonian fluids. The variation of the shear stress with shear strain rate for different types of fluids is shown in Fig. 3.6.

Non-Newtonian fluids can be classified as shear thickening (dilatants) or shear thinning. In the case of the former, the shear stress required to achieve a given shear strain rate increases as the fluid flows. In other words, it becomes increasingly difficult to keep the fluid moving once it starts flowing. Molasses is a good example of a shear thickening fluid. Shear thinning fluids, on the contrary, flow more easily once they start flowing. Blood is a good example of a shear thinning fluid. Shear thinning fluids are also called pseudo-plastic. As the shear thinning behavior becomes more pronounced, the fluid becomes more plastic. In the extreme case, the fluid may yield first before it starts flowing, as in the case of an ideal Bingham plastic. Toothpaste is a good example of such a substance.

The relationship between the shear stress and the shear strain rate for non-Newtonian fluids is usually modelled as

$$\tau_{yx} = 2\kappa\epsilon_{yx}^{n} = \left(2\kappa\epsilon_{xy}^{n-1}\right)\epsilon_{yx} \,, \tag{3.25}$$

where, κ and n have to be obtained from a curve-fit to experimental data. The quantity within the brackets in the second equality is usually called

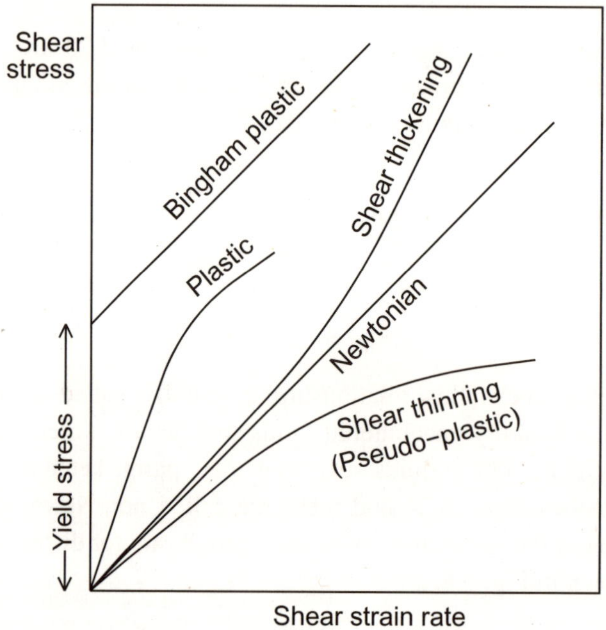

Fig. 3.6: Variation of shear stress and shear strain rate for different types of fluids. Adapted from *Viscous Fluid Flow* by White.

the apparent viscosity. Values of $n < 1$ indicate shear thinning behavior and values of $n > 1$ denote shear thickening behavior and $n = 1$ indicates a Newtonian fluid (with $\kappa = \mu$). There are usually, other difficulties in modeling the relation between the shear stress and the shear strain rate of non-Newtonian fluids. The interested reader can consult textbooks on Rheology for further details. In this book, we will assume Newtonian behavior throughout.

Viscosity arises from the adhesive forces between the molecules in the case of liquids while, in the case of gases, it arises due to the collisions between molecules. The adhesive forces between molecules diminish with increasing temperature and so the viscosity of liquids decrease with temperature. This behavior is of concern in certain applications such as internal combustion engines, since the lubricating oil may not provide the necessary lubricating action at elevated temperatures, especially with prolonged use. The oil thus needs to be replaced frequently.

Viscosity of gases, on the other hand, increases with temperature, owing to the increased molecular kinetic energy and number of collisions between the molecules. However, the viscosity of gases is usually two orders of magnitude lower than that of liquids, which explains the widespread use of liquids as lubricants in many applications. For example, the coefficient of dynamic viscosity for water at 20°C and 1 atm is $1.003 \times 10^{-3}\ N.s/m^2$, while for air, it is $1.8 \times 10^{-5}\ N.s/m^2$. In many applications, where volume is a constraint, the kinematic viscosity, $\nu = \mu/\rho$, rather than the dynamic viscosity, is more relevant. Owing to their low densities, the kinematic viscosity of gases is more than that of liquids. For instance, the kinematic viscosity of water and air at 20°C and 1 atm are $1.005 \times 10^{-6}\ m^2/s$ and $1.5 \times 10^{-5}\ m^2/s$ respectively. It is for this reason that gas filled shock absorbers offer better performance when compared with hydraulic (oil filled) shock absorbers.

3.7 Continuity equation - revisited

In this section, the continuity equation is derived using the Lagrangian approach for the purpose of demonstrating how this approach differs from the Eulerian approach.

Consider the fluid element shown in Fig. 3.1. As this fluid element moves (see Fig. 2.10), the mass conservation principle requires that the time rate of change of the mass of the fluid element be zero. Mathematically,

$$\frac{D}{Dt}(\rho\Delta\mathcal{V}) = 0\,.$$

If we use the product rule to expand this expression, we get

$$\Delta\mathcal{V}\frac{D\rho}{Dt} + \rho\frac{D(\Delta\mathcal{V})}{Dt} = 0\,. \tag{3.26}$$

It was shown in the previous section that the volume of the fluid element changes due to dilatational strains along the coordinate directions. Further, it was shown that the change in the length of the fluid element along the x, y and z directions are, $\frac{\partial u}{\partial x}\Delta x \Delta t$, $\frac{\partial v}{\partial y}\Delta y \Delta t$ and $\frac{\partial w}{\partial z}\Delta z \Delta t$. Hence,

$$\frac{D(\Delta \mathcal{V})}{Dt} = \frac{1}{\Delta t}\left[\left(\Delta x + \frac{\partial u}{\partial x}\Delta x \Delta t\right) \times \left(\Delta y + \frac{\partial v}{\partial y}\Delta y \Delta t\right) \times \left(\Delta z + \frac{\partial w}{\partial z}\Delta z \Delta t\right) - \Delta x \Delta y \Delta z\right].$$

If we expand the brackets, carry through the multiplication and neglect terms involving the products of the velocity gradients, we get

$$\frac{D(\Delta \mathcal{V})}{Dt} = \left(\frac{\partial u}{\partial x} + \frac{\partial v}{\partial y} + \frac{\partial w}{\partial z}\right)\Delta x \Delta y \Delta z\,.$$

Upon substituting this expression into Eqn. 3.26, we are led to

$$\frac{D\rho}{Dt} + \rho\left(\frac{\partial u}{\partial x} + \frac{\partial v}{\partial y} + \frac{\partial w}{\partial z}\right) = 0\,,$$

which is identical to Eqn. 3.10 obtained using the Eulerian approach.

Exercises

(1) The stress tensor at a point P is given as

$$\begin{pmatrix} 6 & -3 & 1 \\ -3 & 2 & 4 \\ 1 & 4 & 8 \end{pmatrix} .$$

Determine the normal and the shear stress vectors on a plane passing through P and parallel to $3x + 4y + 5z = 0$. [(11 19 59)]

(2) Decompose the above stress tensor into its spherical and deviatoric parts and determine the principal deviator stresses, the principal stresses as well as the principal directions.

(3) Consider the velocity field $u = Cx,\ v = -Cy,\ w = 0$. Determine (a) whether this is a valid flow field or not and if valid (b) whether the flow is viscous or inviscid, (c) whether the flow is irrotational or rotational and (d) the components of the rate of strain tensor.

(4) Consider a velocity field (in cylindrical polar coordinates) given by

$$u_r = 0, \quad u_\theta = \frac{\Gamma_0}{2\pi r}\left(1 - e^{-r^2/4\nu t}\right), \quad u_z = 0\,.$$

Determine (a) whether this is a valid flow field or not and if valid, (b) whether the flow is viscous or inviscid and (c) whether the flow is irrotational or rotational.

(5) Consider a velocity field (in cylindrical polar coordinates) given by

$$u_r = U_\infty \cos\theta \left(1 - \frac{a^2}{r^2}\right) \quad \text{and}$$

$$u_\theta = -U_\infty \sin\theta \left(1 + \frac{a^2}{r^2}\right) - \frac{\Gamma}{2\pi r}$$

$$u_z = 0\,.$$

Determine (a) whether this is a valid flow field or not and if valid (b) whether the flow is viscous or inviscid and (c) whether the flow is irrotational or rotational.

Chapter 4

Solutions to the Incompressible Navier-Stokes Equations

In this chapter, some important aspects concerning the mathematical nature of the incompressible Navier-Stokes equations, boundary conditions and the types of solutions are discussed. Many advanced text books that provide a formal mathematical discussion of these topics are available and the interested readers are urged to consult these books.

4.1 Mathematical nature of the incompressible Navier-Stokes equations

Equations 3.11 and 3.23 constitute four equations for the four dependent variables, u, v, w and p. The three momentum conservation equations, 3.23, can be thought of (rightfully so) as the governing equations for each component of the velocity. This leaves the continuity equation, 3.11, as the governing equation for the pressure. However, the pressure, unfortunately, does not appear in the continuity equation at all. The usual practice is to partially differentiate each momentum equation in Eqn. 3.23 with respect to x, y and z respectively and combine the result. This can be accomplished easily if we start from the vector form of Eqn. 3.23, *viz.*,

$$\rho\frac{\partial \vec{u}}{\partial t} + \rho(\vec{u}\cdot\vec{\nabla})\vec{u} = \rho\vec{g} - \vec{\nabla}p + \mu\vec{\nabla}\cdot\vec{\nabla}\vec{u}\,. \tag{4.1}$$

Upon taking the divergence of this equation, we get

$$\rho\frac{\partial}{\partial t}(\vec{\nabla}\cdot\vec{u}) + \rho\left[(\vec{u}\cdot\vec{\nabla})(\vec{\nabla}\cdot\vec{u}) + (\vec{\nabla}\cdot\vec{u})(\vec{\nabla}\cdot\vec{u})\right] = \rho(\vec{\nabla}\cdot\vec{g}) - \vec{\nabla}\cdot\vec{\nabla}p + \mu\vec{\nabla}\cdot\vec{\nabla}(\vec{\nabla}\cdot\vec{u})\,.$$

If we use the fact that $\vec{\nabla}\cdot\vec{u} = 0$, we get

$$\nabla^2 p = 0\,. \tag{4.2}$$

This is the governing equation for the pressure in an incompressible flow.

4.2 Boundary conditions

Equations 3.11 and 3.23 support the trivial solution $u = v = w = 0$ (or, indeed any constant value) with the pressure being equal to thc hydrostatic pressure. For any other situation, appropriate boundary conditions have to be prescribed. In some cases, the boundary condition drives the flow and in other cases, a sourcc term in the governing equations provides the driving force. Since the dependent variables in the equations 4.1 are u, v, w and p, the boundary condition(s) is(are) usually a specification of either the values and/or the derivatives of these quantities. The most common boundary condition is the so-called no-slip boundary condition in which the velocity of the fluid on a solid surface is set equal to the velocity with which the surface itself moves. For instance, for the flow between the parallel plates illustrated in Fig. 2.6(c), the x-component of the velocity on the top and bottom wall would be set to U and zero respectively. The physical basis for this specification arises from the viscosity of the fluid and the nature of the viscous action near solid surfaces. It should, however, be noted that the y-component of velocity of the fluid on the top and bottom wall is equal to zero since the walls are impermeable. This, of course, implies that the walls themselves are streamlines. In addition to the boundary condition for velocity, pressure can also be specified at the inlet and the outlet. In general, it is customary to generate the boundary condition for pressure on each boundary by applying the component of Eqn. 4.1 that is normal to that boundary and then simplifying using the velocity boundary conditions.

For the flow around the square cylinder shown in Figs. 2.2 and 2.3, the x-component of the velocity of the fluid at the inlet is specified to be a non-zero value. The derivative of the x-component of the velocity is specified to be zero on the top and bottom boundaries to simulate the fact that there is no straining of the fluid element due to viscous action in the freestream. The velocity of the fluid on the surface of the cylinder is set to zero on account of the no-slip condition. The fluid is simply allowed to leave at the outlet boundary. The specification of the non-zero value for the velocity at the inlet drives the flow in this case. As we saw earlier, although this value itself is a constant, the resulting flow can be steady or unsteady depending on the actual value.

For the flow shown in Fig. 2.10, the velocity is zero on all the boundaries since the flow is bounded by walls on all sides. The flow in this case is driven by the $\rho\vec{g}$ term owing to the variation in the density from one point in the flow to another.

The specification of the boundary conditions for real problems has to be done with care and should reflect the physics of the problem in order for the solution of the governing equations to be meaningful.

4.3 An illustrative example

Consider the second order ordinary differential equation in the domain $0 \leq y \leq 1$ [†]

$$\epsilon \frac{d^2u}{dy^2} + \frac{du}{dy} = 1\,, \quad \epsilon \ll 1\,, \tag{4.3}$$

with the boundary conditions $u(0) = 0$ and $u(1) = 2$. The parameter ϵ is assumed to be very small. The exact solution to this equation is given as

$$u_{exact} = y + \frac{1 - e^{-y/\epsilon}}{1 - e^{-1/\epsilon}}\,. \tag{4.4}$$

[†] This example has been adapted from the book Asymptotic Analysis by J. D. Murray, Springer-Verlag (1984)

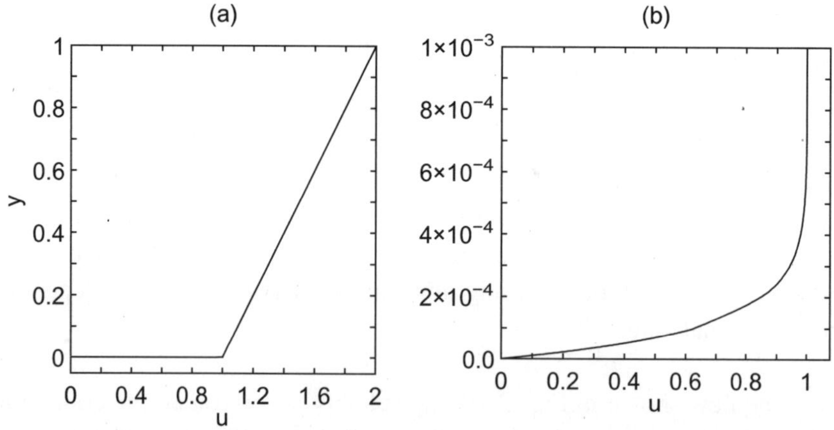

Fig. 4.1: Exact solution for the model equation. (a) Full domain (b) Close-up view near $y = 0$.

The exact solution for $\epsilon = 10^{-4}$ is plotted in Fig. 4.1. It is evident from this figure that the solution has two layers - one for very small values of y in which the variation with y is very rapid and another one in which the variation is linear in y. It will be a very useful exercise to determine ways to construct the solution corresponding to these layers separately and then blending them so as to recover the exact solution. The usefulness of this exercise lies in the fact that, in the general case, it is possible to construct the solutions in the different layers while obtaining the exact solution can be very difficult, if not impossible.

We start by using the fact that $\epsilon \ll 1$ to simplify Eqn. 4.3 thus:

$$\frac{du}{dy} = 1\,. \tag{4.5}$$

This can be solved exactly and the solution is

$$u_{outer} = y + A\,, \tag{4.6}$$

where A is a constant to be determined using the boundary conditions.

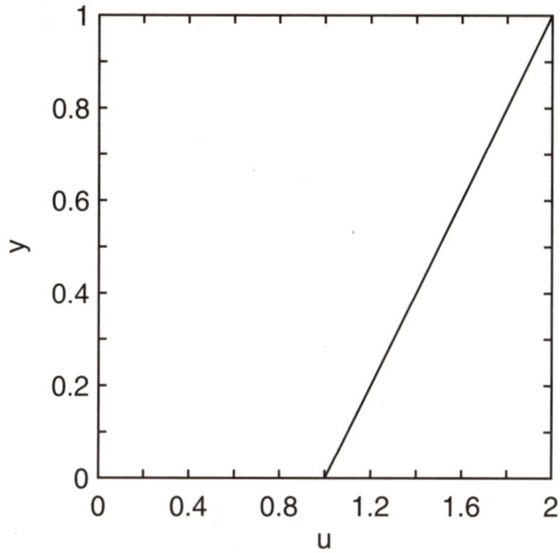

Fig. 4.2: Outer solution for the model equation

Here, we have followed the customary practice of calling this solution the "outer" solution. There is a dilemma in evaluating A since there is only one constant to be evaluated but there are two boundary conditions. This can be resolved by evaluating A using each condition and then trying to take the solution procedure to its logical conclusion with each value of A. Hopefully, we will succeed with one, but not the other value for A. Accordingly, $A = 1$, if we impose $u(1) = 2$ and $A = 0$, if we impose $u(0) = 0$. We proceed with the former choice first and take $u_{outer} = y+1$. This is shown graphically in Fig. 4.2. The outer solution, of course, fails to satisfy the boundary condition at $y = 0$. We construct an "inner" solution that satisfies the condition at $y = 0$ and blends smoothly with the outer solution at some distance away from $y = 0$. In other words, our hypothesis is that the outer solution is acceptable everywhere except in a small region near $y = 0$, where the inner solution is applicable. We are thus faced with two tasks, namely, the determination of the inner solution and the region in which it is applicable. Let us now proceed with the accomplishment of both these tasks.

The dilemma mentioned above in connection with the evaluation of A arises due to the fact that the order of the differential equation reduces from two to one in going from Eqn. 4.3 to 4.5 *i.e.,* when the term $\epsilon d^2u/dy^2$ is neglected. The equation is thus said to be "singular". An examination of Fig. 4.1 shows that d^2u/dy^2 is quite large near $y = 0$ and small everywhere else. Hence, the product $\epsilon \times d^2u/dy^2$ is not negligible near $y = 0$, although ϵ itself is very small. The inner solution has to be determined by taking this into account. Furthermore, it is intuitively obvious that the extent of the inner region must depend in some way on ϵ. It is customary to assume this dependence to be of the form ϵ^m, where m is a positive number that has to be determined along with the inner solution. In essence, we have assumed the thickness of the inner layer to be $O(\epsilon^m)$.

We start by rewriting Eqn. 4.3 using an "inner coordinate" $Y = y/\epsilon^m$, in place of y. Using chain rule, we can write $d/dy = (1/\epsilon^m)d/dY$ and $d^2/dy^2 = (1/\epsilon^{2m})d^2/dY^2$. Equation 4.3 can now be rewritten as

$$\epsilon \frac{1}{\epsilon^{2m}} \frac{d^2u}{dY^2} + \frac{1}{\epsilon^m} \frac{du}{dY} = 1\,.$$

After simplification, this becomes

$$\frac{d^2u}{dY^2} + \epsilon^{m-1} \frac{du}{dY} = \epsilon^{2m-1}\,. \tag{4.7}$$

By equating the sizes of the different terms in this equation, we can obtain different values for m.

(1) If we demand that the terms containing the first and the second derivative be of the same size, then $m = 1$. Equation 4.7 then becomes

$$\frac{d^2u}{dY^2} + \frac{du}{dY} = \epsilon\,.$$

Since $\epsilon \ll 1$, the right hand side of this equation can be set to zero. Thus,

$$\frac{d^2u}{dY^2} + \frac{du}{dY} = 0\,. \tag{4.8}$$

(2) If we demand that the second derivative term be of the same size as the right hand side, then $m = 1/2$. Equation 4.7 then becomes

$$\frac{d^2u}{dY^2} + \frac{1}{\sqrt{\epsilon}}\frac{du}{dY} = 1\,,$$

which can be rewritten as

$$\sqrt{\epsilon}\frac{d^2u}{dY^2} + \frac{du}{dY} = \sqrt{\epsilon}\,.$$

Since $\epsilon \ll 1$, this simplifies to

$$\frac{du}{dY} = 0\,.$$

(3) If we demand that the first derivative term be of the same size as the right hand side, then $m = 0$. Equation 4.7 then becomes

$$\frac{d^2u}{dY^2} + \frac{1}{\epsilon}\frac{du}{dY} = \frac{1}{\epsilon}\,,$$

which can be rewritten as

$$\epsilon\frac{d^2u}{dY^2} + \frac{du}{dY} = 1\,.$$

Since $\epsilon \ll 1$, this simplifies to

$$\frac{du}{dY} = 1\,.$$

It should be noted that the last two choices above lead to a first order equation for the inner solution also. Based on our discussion on the nature of the equation that the inner solution must satisfy, it is easy to discard these choices for the value of m. Hence, the first choice, namely, $m = 1$ is the correct one and Eqn. 4.8 is the equation that the inner solution must satisfy. This choice for the value of m is usually called the "distinguished limit". We note in passing, that, Eqn. 4.8 is second order, in contrast to Eqn. 4.5.

Equation 4.8 requires two boundary conditions. The first, and the most obvious one is $u(y = Y = 0) = 0$. The second one is that the inner solution must blend with the outer solution as Y becomes large *i.e.,* as $Y \to \infty$. In other words, $u_{inner}(Y \to \infty)$ should approach $u_{outer}(y \to 0)$. Since the thickness of the inner layer is $O(\epsilon)$, it might seem reasonable to apply this condition at $y = \epsilon$, or, equivalently, $Y = 1$. However, it must be kept in mind that the development above merely establishes that the thickness of the inner layer is of the order of ϵ and not *equal* to ϵ. For values of y greater than ϵ, Y becomes quite large and hence this condition ensures that the inner solution blends with the outer solution outside the inner layer.

The solution to Eqn. 4.8 subject to the above two boundary conditions is given as

$$u_{inner} = 1 - e^{-Y} . \tag{4.9}$$

This is plotted in Fig. 4.3 for $\epsilon = 10^{-4}$. It is clear from this figure that the thickness of the inner layer is not equal to ϵ, but is about 6ϵ. The blending of the inner solution with the outer solution is shown in Fig. 4.4. It can be seen that for $y > 6\epsilon$, the outer solution begins to depart from the inner solution.

The solution methodology described above is called the singular perturbation method. The interested readers are urged to consult advanced text books on this topic for a more formal development of the concepts and

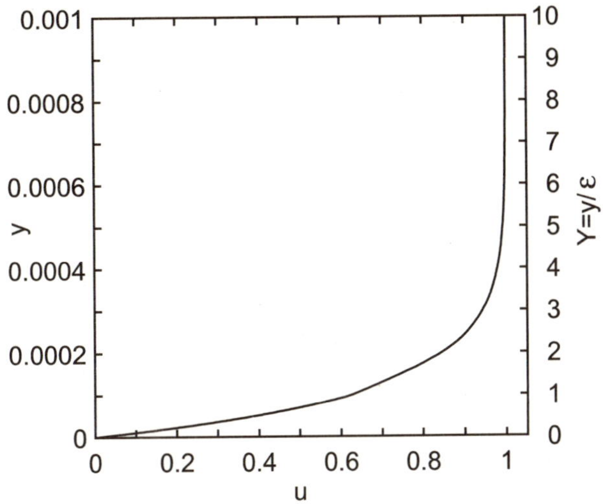

Fig. 4.3: Inner solution for the model equation

ideas†. In the development above, formality has been sacrificed to some extent for the sake of simplicity and ease of exposition.

Finally, by combining Eqns. 4.6 and 4.9, the solution to Eqn. 4.3 with the given boundary conditions obtained using singular perturbation method is

$$u = \begin{cases} y+1, & y \sim O(1)\,, \\ 1-e^{-y/\epsilon}, & y \sim O(\epsilon)\,. \end{cases} \tag{4.10}$$

Comparison of Eqn. 4.10 with 4.4 shows how well the former approximates the latter and this demonstrates the power and usefulness of the singular perturbation technique.

†

(1) J. Kevorkian and J.D. Cole, Perturbation Methods in Applied Mathematics, Springer-Verlag, 1981

(2) J. D. Murray, Asymptotic Analysis, Springer-Verlag, New York, 1984.

(3) A. H. Nayfeh, Perturbation Methods, Wiley Classics Library, 2000.

(4) M. Van Dyke, Perturbation Methods in Fluid Mechanics, Parabolic press, 1975.

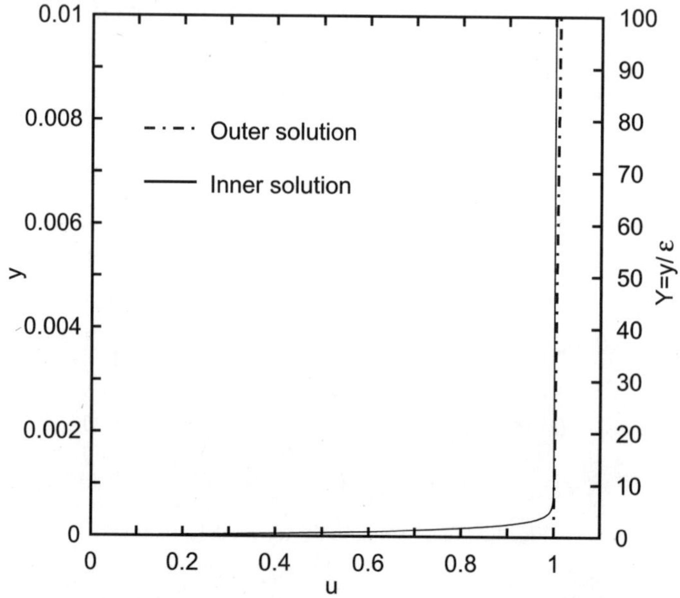

Fig. 4.4: Blending of the inner and the outer solution for the model equation

It may be recalled that there were two choices available for the evaluation of the constant A in the outer solution. By choosing to satisfy the boundary condition at $y = 1$, we were led to $u_{outer} = y + 1$ and subsequently to Eqn. 4.9 near $y = 0$. We now wish to enquire whether the other choice will lead to a consistent outer and an inner solution (near $y = 1$). It is our hope that it will not, as otherwise, we will end up with two seemingly valid but entirely different solutions for the same problem.

If we impose $u = 0$ at $y = 0$ on the outer solution, we get $u_{outer} = y$. This, of course, fails to satisfy the boundary condition at $y = 1$. Proceeding in the same manner as before, we will attempt to construct an inner solution near $y = 1$. The inner coordinate is now defined as $Y = (1 - y)/\epsilon^m$. Using chain rule, we can write $d/dy = -(1/\epsilon^m)d/dY$ and $d^2/dy^2 = (1/\epsilon^{2m})d^2/dY^2$. Eqn. 4.3 can now be rewritten as

$$\epsilon \frac{1}{\epsilon^{2m}} \frac{d^2u}{dY^2} - \frac{1}{\epsilon^m} \frac{du}{dY} = 1 \, .$$

After simplification, this becomes

$$\frac{d^2u}{dY^2} - \epsilon^{m-1}\frac{du}{dY} = \epsilon^{2m-1}\,.$$

It is easy to show that, in this case also, $m = 1$ is the distinguished limit. Thus

$$\frac{d^2u}{dY^2} - \frac{du}{dY} = \epsilon\,.$$

Since $\epsilon \ll 1$, the right hand side of this equation can be set to zero. Thus,

$$\frac{d^2u}{dY^2} - \frac{du}{dY} = 0\,.$$

The boundary conditions for this equation are $u(Y = 0) = 2$ and that $u_{inner}(Y \to \infty) \to u_{outer}(y \to 1)$.

The solution to the above equation is $u_{inner} = C_1 + C_2 e^Y$, where C_1 and C_2 are constants to be determined from the boundary conditions. It is quite obvious that the inner solution cannot be matched with the outer solution as $Y \to \infty$. Thus, it is not possible to construct an inner solution near $y = 1$. This demonstrates convincingly the uniqueness of the solution given by Eqn. 4.10.

An alternative, but faster method of estimating the thickness of the inner layer in such problems is through a scale analysis. It is important to bear in mind that, this method too, yields an estimate of the thickness of the layer if it exists, but does not say whether such a solution is possible or not. We start by assuming that the thickness of the inner layer is δ and the change in u across the inner layer is Δu. The sizes of the terms in Eqn. 4.3 can now be estimated as

$$\epsilon\frac{\Delta u}{\delta^2} \qquad \frac{\Delta u}{\delta} \qquad 1\,.$$

Since the minimum value for u is zero (at $y = 0$) and the maximum value is 2 (at $y = 1$), a good estimate for Δu is 2, the difference between the maximum and the minimum value for u. The above expression then becomes

$$\epsilon\frac{2}{\delta^2} \qquad \frac{2}{\delta} \qquad 1\,.$$

From our earlier discussion, it is clear that the inner equation must contain the second derivative term. Accordingly, if we equate the size of the first and the third term in the above expression, we get $\delta \sim \sqrt{\epsilon}$. Here, the $\sqrt{2}$ factor has been dropped, since we are interested only in an estimate for δ. The first and the third terms are now both $O(1)$, whereas, the second term (which contains the first derivative) is of size $O(1/\sqrt{\epsilon})$, which is much larger. This estimate for δ thus leads us to an equation similar to the outer equation 4.5, which is devoid of the second derivative and hence must be discarded.

On the other hand, if we equate the size of the first and the second term in the above expression, we get $\delta \sim \epsilon$. The first and the second terms are now both $O(1/\epsilon)$, whereas, the third term (the right hand side of Eqn. 4.3) of size $O(1)$, which is much smaller. This will result in an inner equation that will contain the first two terms, with the right hand side set to zero, which is nothing but Eqn. 4.8. The inner coordinate Y is usually defined as $Y = y/\delta$. Scale analysis can thus be seen as a powerful technique and it is widely used in fluid mechanics, heat transfer and in the study of non-linear equations.

4.4 Solutions to the incompressible Navier-Stokes equations

As mentioned in the previous section, the usefulness of the singular perturbation method is that it allows us to construct a composite solution even when it is not possible to obtain an analytical solution. This aspect is desirable especially in the case of the incompressible Navier-Stokes equations, since analytical solutions are not possible except in a few special cases. It is important to keep in mind that it is not always possible to construct a composite, *i.e.,* inner+outer, solution. Since the outer solution (if it exists) is independent of ϵ, a singular perturbation solution can be constructed only if the effect of the smallness parameter ϵ, is confined to a small region or regions. This is brought out clearly in Fig. 4.5.

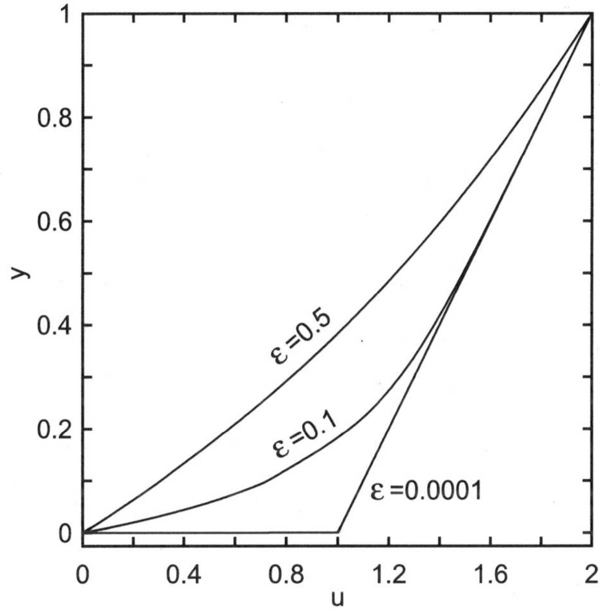

Fig. 4.5: Exact solution for the model equation for different values of ϵ

It is obvious from this figure that the problem admits an inner+outer solution for small values of ϵ, but not for $\epsilon \geq 0.1$. For higher values of ϵ, no clear outer or inner layers are discernable, since the effect of ϵ is not confined to a small inner region. A similar situation arises even for a small value of ϵ, if Eqn. 4.3 were altered slightly to read

$$\epsilon \frac{d^2u}{dy^2} = 1 \,, \quad \epsilon \ll 1 \,, \tag{4.11}$$

with the boundary conditions remaining the same. The exact solution to this problem can be easily determined to be

$$u = \frac{1}{\epsilon} y^2 + \left(2 - \frac{1}{\epsilon}\right) y \, . \tag{4.12}$$

A plot of this solution would show it to be a parabola without any distinctive inner or outer layer.

Comparison of Eqn. 4.1 with Eqn. 4.3 shows that the highest derivative term is indeed multiplied by a smallness parameter, since the coefficient of dynamic viscosity is usually very small. In those types of flows wherein the effect of viscosity is confined to a small layer adjacent to solid surfaces, it should then be possible to construct a singular perturbation solution. The outer solution, obtained by setting μ equal to zero in Eqn. 4.1, is called the inviscid solution. An inner solution, called the boundary layer solution, has to be constructed in the vicinity of solid surfaces to satisfy the no-slip boundary condition on such surfaces. These two solutions are taken up in turn in the next two chapters.

Exercises[†]

In the following problems, $'$ denotes differentiation with respect to x.

(1) Consider the boundary value problem

$$\epsilon u'' - u = x^2 + 1, \qquad u(0) = u(1) = 0\,.$$

Derive the outer solution and the inner solution(s), for $\epsilon \ll 1$.

(2) Consider the boundary value problem

$$\epsilon u'\, u'' + u'^2 - u^2 = 0, \qquad u(0) = 1,\;\; u(1) = 2\,.$$

Derive the outer solution and the inner solution(s), for $\epsilon \ll 1$.

[†] I am grateful to Prof. Foster of The Ohio State University for providing these exercise problems

Chapter 5

Potential Flows

In this chapter, the equations that govern the flow of inviscid fluids are derived from the incompressible Navier-Stokes equations. Simple solutions to these equations are illustrated and methods for constructing more complicated solutions are demonstrated.

5.1 Euler equation for inviscid flows

The equations that govern inviscid flows are obtained by setting the viscosity to zero in the incompressible Navier-Stokes equations 4.1. Thus

$$\nabla \cdot \vec{u} = 0 \tag{5.1}$$

and

$$\rho \frac{\partial \vec{u}}{\partial t} + \rho(\vec{u} \cdot \vec{\nabla})\vec{u} = \rho \vec{g} - \vec{\nabla} p \, . \tag{5.2}$$

Equation 5.2 is called the Euler equation. As alluded to earlier, the order of the governing equations has decreased as a consequence of this approximation. As a result, the number of boundary conditions that can be imposed also decreases. The exact conditions that can be imposed are problem dependent, but some general comments can be made. In particular, the specification of boundary conditions on solid surfaces requires special consideration. It may be recalled that the no-slip condition requires the tangential components of velocity on the surface be equal to the velocity of the surface itself and the normal component of velocity on the surface is set to zero if the surface is impermeable. It can be seen that imposing the former condition would require the generation of shear stresses along the surface *i.e.,* non-zero gradient of the tangential components of the velocities along the normal direction to the surface. However, Eqn. 3.21 shows that $\bar{\bar{\Pi}}' = 0$ for an inviscid fluid and so $\bar{\bar{\Pi}} = -p\,\bar{\bar{I}}$. In other words, all the off-diagonal components of the stress tensor which represent shear stresses are zero. Hence, it is physically not possible to impose the no-slip condition on solid surfaces in inviscid flows and so it has to be dropped. The impermeability condition alone can be imposed. In fact, since the off-diagonal components of the stress tensor are zero, it is not possible to impose *any* boundary condition on the tangential velocities or their derivatives.

5.1.1 *Bernoulli's equation*†

The Euler equation 5.2 is written in Cartesian coordinates. A more useful form can be obtained by writing it in streamline coordinates instead of Cartesian coordinates. Before we proceed to derive this form of the equation, some preliminary details have to be worked out.

We start by making the following observation regarding the gradient vector, $\vec{\nabla}$. When it operates on a scalar quantity ϕ, we get a vector whose components are the derivatives of ϕ along the cartesian coordinate directions, *i.e.,* $\vec{\nabla}\phi = \frac{\partial \phi}{\partial x}\hat{\imath} + \frac{\partial \phi}{\partial y}\hat{\jmath} + \frac{\partial \phi}{\partial z}\hat{z}$. When we take the dot product of this quantity with a vector $\vec{dr} = dx\,\hat{\imath} + dy\,\hat{\jmath} + dz\,\hat{k}$, we get a scalar quantity

$$\vec{\nabla}\phi \cdot \vec{dr} = \frac{\partial \phi}{\partial r} = \frac{\partial \phi}{\partial x}\,dx + \frac{\partial \phi}{\partial y}\,dy + \frac{\partial \phi}{\partial z}\,dz\,,$$

which is nothing but the total derivative $d\phi$ along the direction $\vec{dr}$.

Now consider the streamlines sketched in Fig. 5.1. The unit vectors $\hat{e}_s$ and $\hat{e}_n$, along and normal to the streamlines are also shown in this figure. When

†I am grateful to Prof. Korpela of the Ohio State University for graciously providing some of the material in this subsection

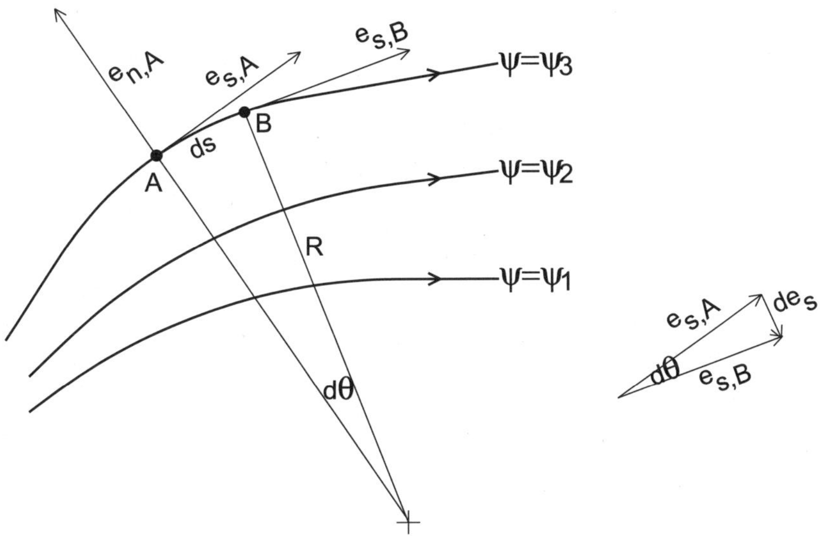

Fig. 5.1: Illustration of streamline coordinates

the unit tangent vector $\hat{e}_s$ is displaced along the streamline from A to B, there is also a rotation of the vector by $d\theta$. This displacement is indicated as $d\hat{e}_s$ in Fig. 5.1. It should also be noted that this displacement is along the negative $\hat{e}_n$ direction (see inset in Fig. 5.1). For small displacements, it is easy to see that

$$d\hat{e}_s = -\frac{ds}{R}\hat{e}_n \, , \tag{5.3}$$

where R is the radius of curvature of the streamline.

We proceed now with the derivation of Eqn. 5.2 in streamline coordinates. This can be accomplished by taking the dot product with $\hat{e}_s$. Let us now do this term by term.

$$\rho\vec{g} \cdot \hat{e}_s = -\rho g \hat{k} \cdot \hat{e}_s = -\rho g \vec{\nabla} z \cdot \hat{e}_s = -\rho g \frac{\partial z}{\partial s} \, , \tag{5.4}$$

where, we have assumed that $\vec{k}$ points vertically upward. Next,

$$-\vec{\nabla}p \cdot \hat{e}_s = -\frac{\partial p}{\partial s} \,. \tag{5.5}$$

Since $\vec{u}$ and $\hat{e}_s$ are parallel, we can write $\vec{u} = V\hat{e}_s$, where V is the velocity magnitude. The convective term in Eqn. 5.2 can be written as

$$\rho(\vec{u} \cdot \vec{\nabla})\vec{u} = \rho V(\hat{e}_s \cdot \vec{\nabla})(V\hat{e}_s) \,.$$

As discussed earlier, $\hat{e}_s \cdot \vec{\nabla}$ is the derivative along the direction of the streamline. Therefore,

$$\begin{aligned} \rho(\vec{u} \cdot \vec{\nabla})\vec{u} &= \rho V \frac{\partial}{\partial s}(V\hat{e}_s) \\ &= \rho V \left(\frac{\partial V}{\partial s}\hat{e}_s + V\frac{\partial \hat{e}_s}{\partial s} \right) \\ &= \rho V \left(\frac{\partial V}{\partial s}\hat{e}_s - \frac{V}{R}\hat{e}_n \right) , \end{aligned}$$

where we have used Eqn. 5.3. If we now take the dot product with $\hat{e}_s$, we get

$$\left[\rho(\vec{u} \cdot \vec{\nabla})\vec{u}\right] \cdot \hat{e}_s = \rho V \frac{\partial V}{\partial s} = \rho \frac{1}{2} \frac{\partial (V^2)}{\partial s} \,. \tag{5.6}$$

Lastly,

$$\rho \frac{\partial \vec{u}}{\partial t} \cdot \hat{e}_s = \rho \frac{\partial (V\hat{e}_s)}{\partial t} \cdot \hat{e}_s = \rho \frac{\partial V}{\partial t} \,. \tag{5.7}$$

Putting the last four equations together, we get

$$\rho \frac{\partial V}{\partial t} + \rho \frac{1}{2} \frac{\partial (V^2)}{\partial s} = -\frac{\partial p}{\partial s} - \rho g \frac{\partial z}{\partial s} \,.$$

Upon integrating this equation along a streamline passing through two points 1 and 2, we get

$$\int_1^2 \frac{\partial V}{\partial t}\, ds + \frac{p_2 - p_1}{\rho} + g(z_2 - z_1) + \frac{V_2^2 - V_1^2}{2} = 0\,. \tag{5.8}$$

This equation is called the unsteady Bernoulli equation. If the flow is steady, the first term goes to zero and we are led to

$$\frac{p_2 - p_1}{\rho} + g(z_2 - z_1) + \frac{V_2^2 - V_1^2}{2} = 0\,. \tag{5.9}$$

Alternatively,

$$\frac{p_2}{\rho} + gz_2 + \frac{V_2^2}{2} = \frac{p_1}{\rho} + gz_1 + \frac{V_1^2}{2} = constant \text{ (along the streamline)}\,. \tag{5.10}$$

If velocity and pressure changes take place only along streamlines and not across, then one streamline is no different from another and so we can write

$$\frac{p}{\rho} + gz + \frac{V^2}{2} = constant \text{ (everywhere)}\,. \tag{5.11}$$

Equation 5.11 can be derived directly from Eqn. 5.2 by requiring that the vorticity $\vec{\nabla} \times \vec{u}$ be zero everywhere. We start with the steady form of the Euler equation,

$$(\vec{u} \cdot \vec{\nabla})\vec{u} = \vec{g} - \frac{1}{\rho}\vec{\nabla}p\,. \tag{5.12}$$

If we rewrite the left hand side using a well known vector identity, we get

$$\frac{1}{2}\vec{\nabla}(\vec{u} \cdot \vec{u}) - \vec{u} \times (\vec{\nabla} \times \vec{u}) = \vec{g} - \frac{1}{\rho}\vec{\nabla}p\,.$$

If we set $\vec{\nabla} \times \vec{u} = 0$ and write $\vec{g} = -g\,\hat{k} = -g\vec{\nabla}z$, we get

$$\frac{1}{2}\vec{\nabla}(\vec{u} \cdot \vec{u}) + \frac{1}{\rho}\vec{\nabla}p + g\vec{\nabla}z = 0\,.$$

Or

$$\vec{\nabla}\left(\frac{1}{2}V^2 + \frac{p}{\rho} + gz\right) = 0\,.$$

It follows that

$$\frac{p}{\rho} + gz + \frac{V^2}{2} = constant\,,$$

which is the same as Eqn. 5.11.

Alternative forms of Eqn. 5.11 are also useful. For example, with a simple rearrangement, we can get

$$p + \rho gz + \frac{\rho V^2}{2} = constant\,. \tag{5.13}$$

Note that all the terms have dimensions of N/m^2. The first term is of course, the pressure, the second term is the hydrostatic pressure and the third term is usually called the dynamic pressure. Yet another rearrangement of Eqn. 5.11 gives

$$\frac{p}{\rho g} + z + \frac{V^2}{2g} = constant\,. \tag{5.14}$$

Now all the terms have dimensions of m. The first term is called the pressure head, the second term is the gravity head and the third term is called the velocity head .

Example 5.1. Consider a syphon that drains the liquid from a large tank as shown in Fig. 5.2. Determine the exit velocity and the pressure at point A.

Solution. Since the tank is large, the level of the liquid remains the same and the operation can be taken to be steady. We will apply the steady Bernoulli equation 5.9 along the streamline shown as a dashed line in Fig.

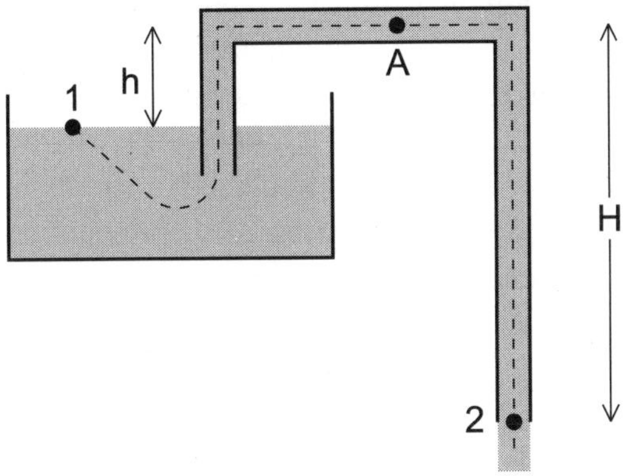

Fig. 5.2: Analysis of a syphon using the Bernoulli equation

5.2 between points 1 and 2. Thus

$$\frac{p_1}{\rho} + gz_1 + \frac{V_1^2}{2} = \frac{p_2}{\rho} + gz_2 + \frac{V_2^2}{2} \, .$$

If we set $p_1 = p_2 = p_{atm}$ and $V_1 = 0$ (since the liquid level does not change), we get

$$V_2 = \sqrt{2g(z_1 - z_2)} = \sqrt{2g(H - h)} \, .$$

If we apply steady Bernoulli equation between points 1 and A, we get

$$\frac{p_1}{\rho} + gz_1 + \frac{V_1^2}{2} = \frac{p_A}{\rho} + gz_A + \frac{V_A^2}{2} \, .$$

If we set $p_1 = p_{atm}$, $V_1 = 0$ and $V_a = V_2$, we get

$$p_A = p_{atm} + g(z_1 - z_A) - \frac{V_2^2}{2} = p_{atm} - gh - g(H - h) = p_{atm} - gH \, .$$

It should be noted that $p_A < p_{atm}$. Also, we have used the fact that point

A is at a higher elevation than point 1. As one moves along the streamline, the pressure increases from p_{atm} to a higher value at entry into the tube, then increases again to p_{atm} at the exit to the tube. The velocity of the fluid inside the tube is equal to $\sqrt{2g(H-h)}$, but it is zero in the tank. In reality, the change in velocity is not discontinuous, but occurs in a small region near the tube entrance. The crowding of the streamlines near the tube entrance causes tremendous acceleration of the fluid, which is neglected in this simplified analysis.

Since the flow is inviscid and irrotational, Eqn. 5.11 can also be applied between any two points in the flow, without regard to the streamlines. □

Example 5.2. Consider the tank shown in Fig. 5.3 with a nozzle attached to it near the bottom. The area of the tank is A_t and the exit area of the nozzle is A_j. Initially, the water level in the tank is at a height h_0 from the nozzle centerline. The nozzle is now opened. Determine the velocity of the jet at the nozzle exit if (a) $A_t \gg A_j$ and (b) A_t/A_j is finite.

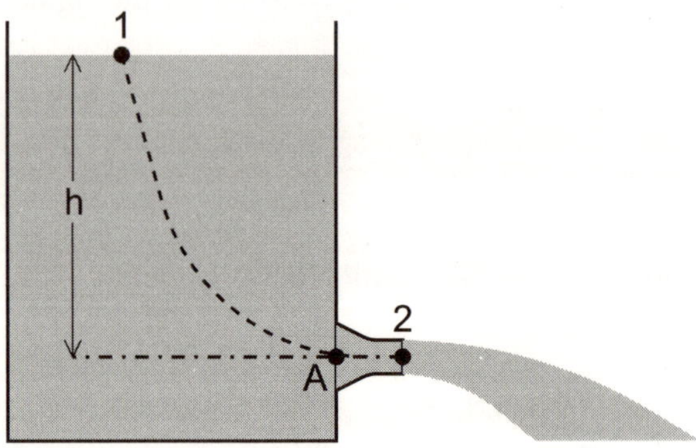

Fig. 5.3: Analysis of the emptying of a tank using the Bernoulli equation

Solution. (a) When $A_t \gg A_j$, we can assume that the water level in the tank does not change with time. If we apply the steady Bernoulli equation 5.11 between points 1 and 2, and note that $p_1 = p_2 = p_{atm}$, $V_1 = 0$, we get

$$V_2 = \sqrt{2g(z_1 - z_2)} = \sqrt{2gh_0}\,.$$

This is the famous Torricelli's formula. Note that, the change in the fluid velocity from the nozzle inlet to the exit is ignored in this analysis.

(b) When A_t/A_j is finite, then the flow is unsteady, since the water level in the tank changes with time. At any instant, let the height of the water level in the tank from the nozzle centerline be h. If we apply the unsteady Bernoulli equation 5.8 along the streamline connecting points 1 and 2 (shown dashed in the figure), we get

$$\int_1^2 \frac{\partial V}{\partial t}\,ds - gh + \frac{V_2^2 - V_1^2}{2} = 0\,,$$

where we have used $p_1 = p_2 = p_{atm}$. Also, $V_2 = (A_t/A_j)V_1$ since the rate of decrease of liquid level in the tank is due to the efflux of mass through the nozzle. Therefore, the above equation becomes,

$$\int_1^2 \frac{\partial V}{\partial t}\,ds - gh + \frac{V_2^2}{2}\left[1 - \left(\frac{A_j}{A_t}\right)^2\right] = 0\,.$$

Point A in Fig. 5.3 also lies on the streamline connecting points 1 and 2. Hence, the integral in the above equation can be split into two parts and this gives

$$\int_1^A \frac{\partial V}{\partial t}\,ds + \int_A^2 \frac{\partial V}{\partial t}\,ds - gh + \frac{V_2^2}{2}\left[1 - \left(\frac{A_j}{A_t}\right)^2\right] = 0\,.$$

The velocity of the fluid in the tank is assumed to be the same (and equal to V_1) everywhere and as mentioned earlier, the velocity of the fluid in the nozzle is taken to be V_2. Hence, there is a discontinuous change in the velocity (and pressure) across point A. Thus

$$\int_1^A \frac{\partial V}{\partial t}\,ds = \frac{dV_1}{dt}\int_1^A ds \approx \frac{dV_1}{dt}\,h\,,$$

and

$$\int_A^2 \frac{\partial V}{\partial t}\, ds = \frac{dV_2}{dt}\, l\,,$$

where l is the length of the nozzle. If we assume $l \ll h$, then the second part of the integral can be neglected. Thus

$$\frac{dV_1}{dt}\, h - gh + \frac{V_2^2}{2}\left[1 - \left(\frac{A_j}{A_t}\right)^2\right] = 0\,.$$

If we set $V_1 = (A_j/A_t)V_2$, we get

$$\frac{A_j}{A_t}\frac{dV_2}{dt}\, h - gh + \frac{V_2^2}{2}\left[1 - \left(\frac{A_j}{A_t}\right)^2\right] = 0\,.$$

We note that, if $A_j/A_t \ll 1$, then the above equation reduces to $V_2 = \sqrt{2gh}$, as it should. Since $V_1 = -dh/dt$ (as h decreases with time, the negative sign ensures that the velocity V_1 is positive), $V_2 = -(A_t/A_j)dh/dt$. We are thus finally led to

$$h\frac{d^2h}{dt^2} - \frac{1}{2}\left(\frac{dh}{dt}\right)^2\left(\frac{A_t}{A_j}\right)^2\left[1 - \left(\frac{A_j}{A_t}\right)^2\right] + gh = 0\,,$$

subject to the conditions $h(t = 0) = h_0$ and $h'(t = 0) = -\sqrt{2gh_0}$. Numerical solutions to this equation for three values of A_t/A_j are shown in Fig. 5.4. As expected, the smaller the tank (for a given nozzle area), the faster it drains. □

Equation 5.9 was derived by taking the dot product of the Euler equation with $\hat{e}_s$. A similar exercise can be carried out by taking the dot product with $\hat{e}_n$. For the sake of simplicity, if we assume the flow to be steady, and then take the dot product, we have

$$\left[\rho(\vec{u}\cdot\vec{\nabla})\vec{u}\right]\cdot\hat{e}_n = -\vec{\nabla}p\cdot\hat{e}_n - \rho g\vec{\nabla}z\cdot\hat{e}_n\,.$$

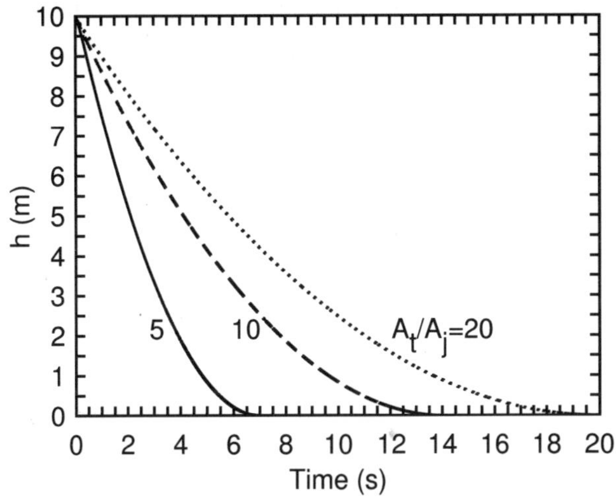

Fig. 5.4: Variation of the height of the fluid level with time during emptying

If we rewrite the left hand side as before,

$$\rho V \left(\frac{\partial V}{\partial s} \hat{e}_s - \frac{V}{R} \hat{e}_n \right) \cdot \hat{e}_n = -\frac{\partial p}{\partial n} - \rho g \frac{\partial z}{\partial n} .$$

Hence

$$\frac{\partial p}{\partial n} = \rho \frac{V^2}{R} - \rho g \frac{\partial z}{\partial n} .$$

If there is no variation in the elevation along the direction normal to the streamlines, then

$$\frac{\partial p}{\partial n} = \rho \frac{V^2}{R} . \tag{5.15}$$

Equation 5.15 shows that the pressure increases in a radially outward direction (see Fig. 5.1). If the flow has to negotiate a sharp corner (for which $R = 0$), then, as Eqn. 5.15 makes clear, the velocity magnitude V

has to go to zero at the corner for the pressure force (and hence the normal acceleration) to be finite.

Example 5.3. Consider the tank shown in Fig. 5.5 which is filled with a fluid and stationary initially. The tank is then made to rotate about its own axis with an angular velocity of Ω. Determine the shape of the free surface.

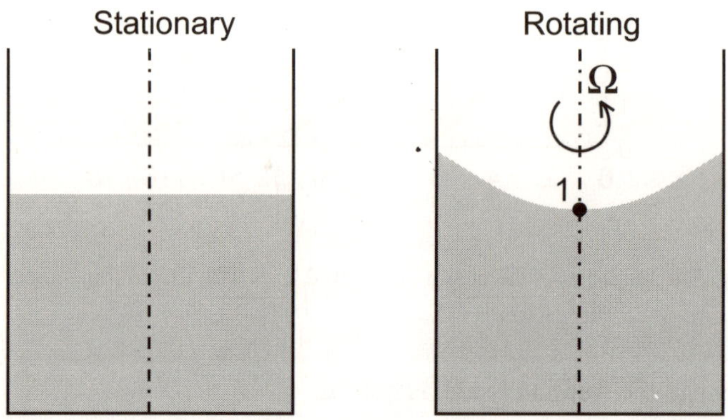

Fig. 5.5: Analysis of solid body rotation using the Bernoulli equation

Solution. When the container is rotated about its own axis, the fluid inside undergoes a solid body rotation. The streamlines for this flow are circular (when viewed from above) as shown in Fig. 2.6(a). Also, note that the vorticity in this case is non-zero as demonstrated in Chapter 2 and so a direct application of Bernoulli's equation, while possible, does not lead to a solution.

Equation 5.15 is applicable since there is no change in the elevation along the radial direction (normal to the streamlines). Thus

$$\frac{\partial p}{\partial r} = \rho \frac{V^2}{r} = \rho r \Omega^2 \,,$$

where we have used the fact that $V = r\Omega$. It is clear from this equation that the pressure increases from the axis towards the container wall. Hence the fluid is pushed outwards towards the wall and the liquid level falls near the axis and increases near the wall. Upon integrating this equation, we get

$$p = \rho \frac{r^2 \Omega^2}{2} + f(z) + constant\,,$$

where $f(z)$ is an arbitrary function in z. It is not very difficult to guess this function. Thus

$$p = \rho \frac{r^2 \Omega^2}{2} - \rho g z + constant\,,$$

where z is the elevation of the point from the bottom of the container. Since $p = p_{atm}$ at point 1, the constant in this equation can be evaluated. Thus

$$p = \rho \frac{r^2 \Omega^2}{2} - \rho g (z - h_1) + p_{atm}\,.$$

The height of the liquid surface h as a function of r and z can be obtained simply by setting $p = p_{atm}$,

$$h - h_1 = \frac{r^2 \Omega^2}{2g}\,.$$

This is the equation of a parabola and hence the liquid surface is a paraboloid. □

5.1.2 *Relation between Bernoulli's equation and the first law of thermodynamics*

Although the steady Bernoulli equation 5.9 has been derived from the Euler equation 5.2, which is a statement of momentum balance, it can also be derived from the first law of thermodynamics. The first law of thermodynamics for a steady flow system with a single inlet and outlet can be written as (consult any standard text book on Thermodynamics)

$$0 = \dot{Q} - \dot{W} + \dot{m}\left(h_1 + \frac{V_1^2}{2} + gz_1 - h_2 - \frac{V_2^2}{2} - gz_2\right). \tag{5.16}$$

The change in enthalpy is given as $dh = d(u+pv) = du+pdv+vdp$. In the absence of any change in temperature, $du = 0$ and for an incompressible fluid, $dv = 0$. Hence, $dh = vdp = dp/\rho$. In addition, if there is no heat or work interaction, Eqn. 5.16 reduces to the steady Bernoulli equation 5.9.

Although the inviscid approximation seems to be very restrictive, Bernoulli's equation can be used in a large number of flows to get reasonably good answers. Heat addition, work addition (as in a fan or pump), density change can all be modelled as discontinuous changes and included in Eqn. 5.9. Loss of pressure in pipes and ducts due to viscous effect, can be modelled as a head loss and included in Eqn. 5.9.

It is clear from the above development that in addition to the simplification offered by the inviscid approximation, further simplifications are possible if the flow is irrotational as well. We turn to a study of such flows next.

5.2 Potential flows

Flows of an incompressible, inviscid fluid, in which the vorticity is zero everywhere are called potential flows. The only exception is that isolated singularities, such as the free vortex discussed in Chapter 2 (it may be recalled that the flow associated with this vortex is irrotational everywhere except at the center), are allowed. Since $\vec{\omega} = \vec{\nabla} \times \vec{u} = 0$, we can write

$$\vec{u} = \vec{\nabla}\phi\,, \tag{5.17}$$

where ϕ is a scalar function called the velocity potential. Since $\nabla \cdot \vec{u} = 0$ (Eqn. 5.1),

$$\nabla^2\phi = 0\,. \tag{5.18}$$

This is the Laplace equation for the velocity potential and is well known in many branches of science and engineering. Consequently, many methods are available for its solution. Note that Eqns. 5.1 and 5.2 have two unknowns - the velocity and the pressure. Once the velocity potential is known, the velocity can be calculated from Eqn. 5.17. The pressure can then be calculated from Bernoulli equation 5.8 for unsteady flows or Eqn. 5.9 for steady flows.

One question that naturally arises at this point is, since Eqn. 5.18 (and indeed Eqn. 5.1) is a steady equation, how can the time dependence of the flow be determined? The answer to this question is that any time dependent behavior of the flow field has to come solely from the boundary conditions to Eqn. 5.18. Also, since Eqn. 5.18 does not contain any time derivative, the transient behavior of the flow field is *identical* to the unsteadiness imposed through the boundary condition. In other words, the flow responds instantaneously to the external forcing through the boundary conditions.

It is important to note that, Eqn. 5.18 alone is solved. Velocity and pressure are *calculated - not solved* from Eqn. 5.17 and the Bernoulli equation. This calculation is made easy by the fact that Eqn. 5.17 and the Bernoulli equation are linear in velocity and pressure respectively. Since Eqn. 5.18 is linear, solutions to it can be superposed to obtain new solutions. For instance, if ϕ_I and ϕ_{II} are solutions to Eqn. 5.18, then $\phi_{III} = \phi_I + \phi_{II}$ is also a solution, since

$$\nabla^2 \phi_{III} = \nabla^2 \phi_I + \nabla^2 \phi_{II} = 0 \, .$$

This suggests that, by starting from solutions to basic flows (which are easily obtained), complicated solutions can be obtained simply by superposition. These solutions then have to be interpreted to determine the actual flow situation they represent. This is demonstrated in the the following sections.

It should be noted that the velocity field can also obtained by superposition, since

$$\vec{u}_{III} = \vec{\nabla} \phi_{III} = \vec{\nabla} \phi_I + \vec{\nabla} \phi_{II} = \vec{u}_I + \vec{u}_{II} \, .$$

The pressure field, however, cannot be obtained by superposition, since the Bernoulli equation contains a non-linear term, V^2. Since $V_{III}^2 = (V_I + V_{II})^2 \neq V_I^2 + V_{II}^2$, it follows that $p_{III} \neq p_I + p_{II}$.

For 2D flows, contours of streamfunction (which are the streamlines) can be calculated from Eqn. 2.6 as

$$\left(\frac{dy}{dx}\right)_{\psi=const} = \frac{v}{u}\,.$$

Since $d\phi = \dfrac{\partial \phi}{\partial x}\, dx + \dfrac{\partial \phi}{\partial y}\, dy$, equi-potential ($\phi = constant$) lines can be determined by setting $d\phi = 0$ in this equation. This gives

$$\left(\frac{dy}{dx}\right)_{\phi=const} = -\frac{\dfrac{\partial \phi}{\partial x}}{\dfrac{\partial \phi}{\partial y}} = -\frac{u}{v}\,,$$

after using Eqn. 5.17. Upon combining the last two equations, we get

$$\left(\frac{dy}{dx}\right)_{\psi=const} \times \left(\frac{dy}{dx}\right)_{\phi=const} = -1\,,$$

which demonstrates that the equi-potential lines are everywhere orthogonal to the streamlines.

5.2.1 *Basic flows*

The four basic flows in 2D (plane) potential flows are shown in Fig. 5.6. The streamlines are shown as solid lines and the potential lines are shown as dashed lines. For the uniform flow at an angle of attack, it is easy to show that

$$\begin{aligned} \phi &= U_\infty x \cos\alpha + U_\infty y \sin\alpha\,, \\ u &= U_\infty \cos\alpha\,; \qquad v = U_\infty \sin\alpha\,, \\ \psi &= U_\infty y \cos\alpha - U_\infty x \sin\alpha\,, \end{aligned} \tag{5.19}$$

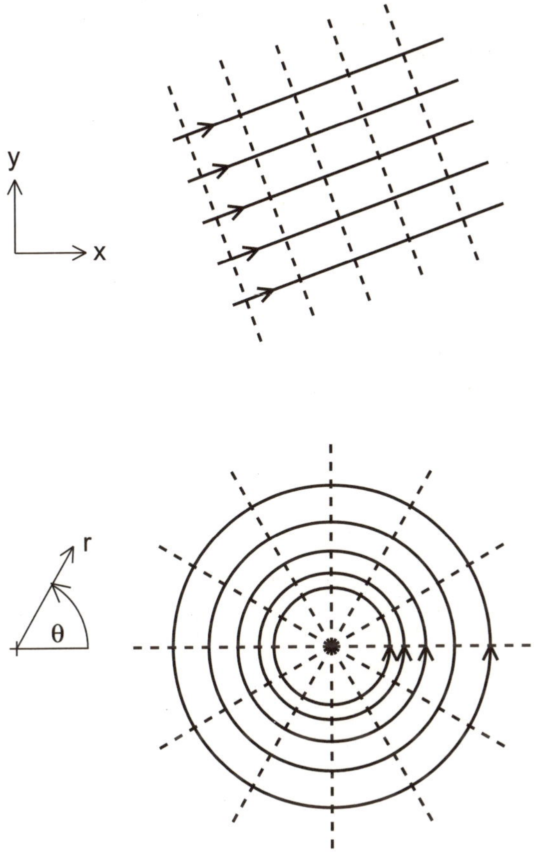

Fig. 5.6: Basic flows (a) Uniform flow at an angle of attack and (b) Vortex flow

where, U_∞ is the freestream velocity and α is the angle of attack with respect to the horizontal. For the flow induced by a vortex[†] with its axis normal to the page and located at the origin,

$$\begin{aligned} \phi &= \frac{\Gamma}{2\pi}\,\theta\,, \\ u_r &= 0\,; \qquad u_\theta = \frac{\Gamma}{2\pi r}\,, \\ \psi &= -\frac{\Gamma}{2\pi}\ln r\,, \end{aligned} \tag{5.20}$$

[†]The vortex is shown in this figure to be in the counter-clockwise direction for the sake of illustration only.

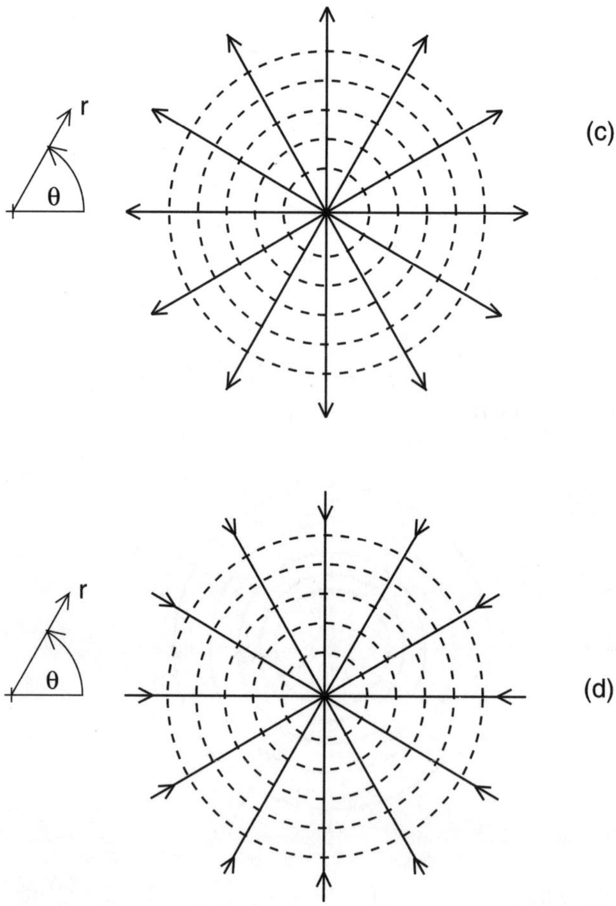

Fig. 5.6: *(cont'd)* Basic flows (c) Source flow and (d) Sink flow

where Γ is the circulation around any curve enclosing the origin. It is important to note that the flow induced by the vortex is irrotational everywhere, except at the origin.

For the flow due to a source located at the origin,

$$\phi = \frac{q}{2\pi} \ln r \,,$$

$$u_r = \frac{q}{2\pi r} \,; \qquad u_\theta = 0 \,, \tag{5.21}$$

$$\psi = \frac{q}{2\pi} \theta \,,$$

where q is the volumetric flow rate per unit length (also called the strength). It must be borne in mind that since this is a plane (2D) source, it is actually a line source that extends to infinity along the direction(s) normal to the page. This is the reason why q is the volumetric flow rate per unit length and not volumetric flow rate itself. If we imagine a cylinder centered about the line source, the volume flow rate that crosses the cylindrical surface equals q.

For the flow due to a sink located at the origin, the same expressions as those for a source apply, but with a change in the sign. Thus

$$\phi = -\frac{q}{2\pi} \ln r \,,$$

$$u_r = -\frac{q}{2\pi r} \,; \qquad u_\theta = 0 \,, \tag{5.22}$$

$$\psi = -\frac{q}{2\pi} \theta \,.$$

5.2.2 *Superposed flows*

Consider the situation when a source and sink of equal strength are located on the x-axis at $x = -a$ and a respectively as shown in Fig. 5.7. The superposed flow field can be obtained by combining Eqns. 5.21 and 5.22 appropriately† as follows:

$$\psi = \frac{q}{2\pi} \left(\theta_I - \theta_{II} \right) \,, \tag{5.23}$$

†Equations 5.21 and 5.22 are written for the case when the source (and sink) are located at the origin. However, now, the source and sink are located at $x = \mp a$ respectively and hence the r and θ in Eqns. 5.21 and 5.22 have to be measured with respect to the source and the sink and not the origin.

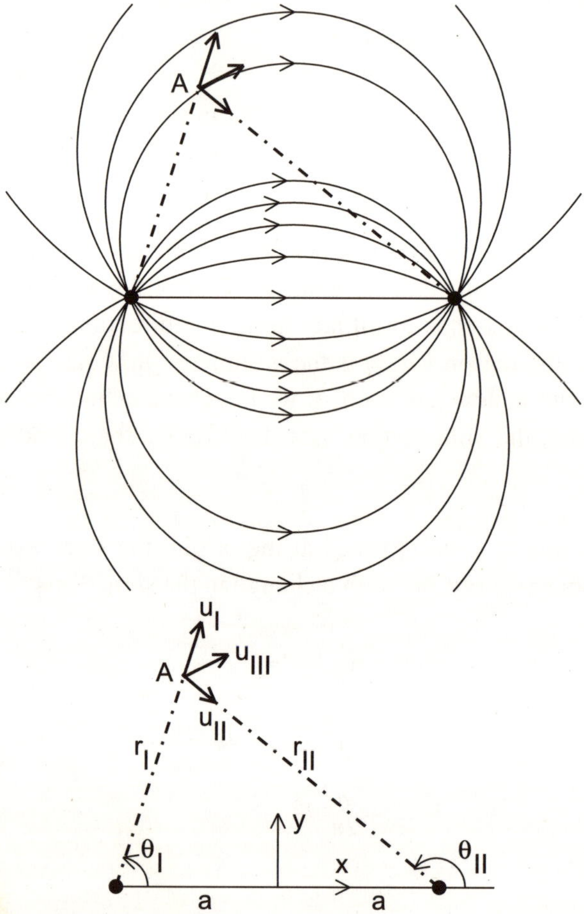

Fig. 5.7: Superposition of a source and sink of equal strength located a distance $2a$ apart on the x-axis

where, subscripts I and II denote that the angles are measured with respect to the source and sink respectively. The resultant velocity field at a point A is illustrated in Fig. 5.7. Note that, the velocity component at A due to the source is more than that due to the sink, since A is closer to the source. If we write the above equation in Cartesian coordinates, we get

$$\psi = -q \tan^{-1}\left(\frac{2ay}{x^2 + y^2 - a^2}\right). \tag{5.24}$$

The velocity components can be calculated from this expression using Eqn. 2.4.

If we now superpose a uniform stream at zero angle of attack on the

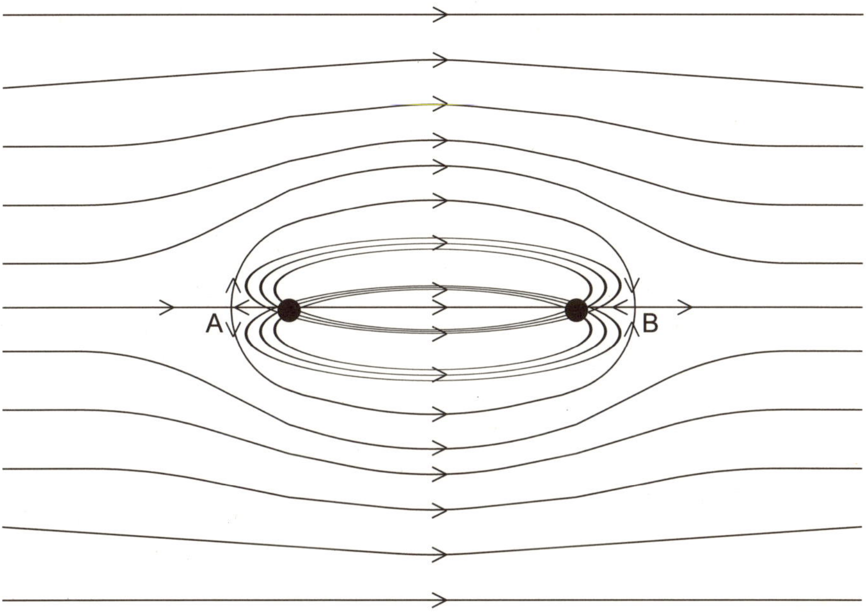

Fig. 5.8: Superposition of an uniform flow, source and sink of equal strength located a distance $2a$ apart on the x-axis

flow shown in Fig. 5.7, the resultant flow is sketched in Fig. 5.8. The streamfunction for this case is given as

$$\psi = -q\tan^{-1}\left(\frac{2ay}{x^2+y^2-a^2}\right) + U_\infty y\,, \tag{5.25}$$

where, U_∞ is the freestream velocity. The velocity is identically zero at points A and B in Fig. 5.8 and these are called stagnation points. Two streamlines - one from the source and one from the freestream, intersect at point A. In addition, two streamlines branch off vertically upwards and downwards. A similar situation exists at point B. It may be recalled from

Chapter 2 that streamlines can intersect only points of zero velocity. If we apply Bernoulli equation along the streamline that approaches point A from the freestream, we get

$$p_A = p_\infty + \frac{\rho U_\infty^2}{2} \, . \tag{5.26}$$

The pressure p_A at the stagnation point is called the stagnation pressure and it is clear from this equation that the deceleration of the flow along this streamline is accompanied by an increase in the pressure. Of course, it is easy to see that $p_B = p_A$ in this case.

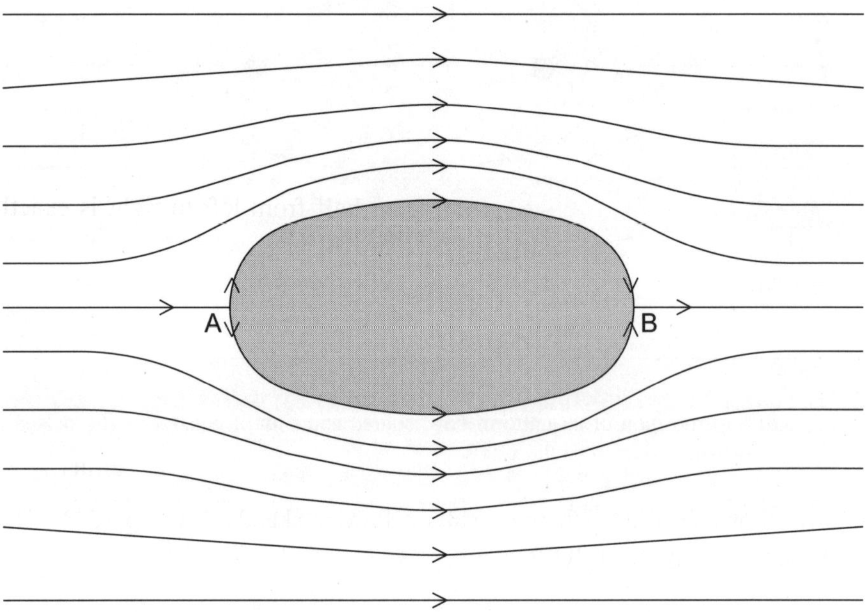

Fig. 5.9: Flow around a Rankine body

The streamlines connecting points A and B form a closed loop and can be seen to completely enclose and isolate the flow inside from that on the outside. Hence, this streamline can actually be thought of as an impermeable surface. Accordingly, if we replace this streamline with an impermeable surface[†] and neglect the flow inside, then Eqn. 5.25 describes the flow around a blunt nosed body (called the Rankine body) as illustrated in Fig. 5.9. The dimensions of the Rankine body are determined by the strength of the source (or sink) and the freestream velocity. The distance between the source and the front stagnation point (as well as the the sink and the rear stagnation point) is equal to q/U_∞. The length of the Rankine body is thus $2a+q/U_\infty$. An increase in the freestream velocity keeping the strength of the source and sink the same, reduces the size of the Rankine body and *vice versa*. It should also be noted that the flow slips along the surface of the Rankine body, as it should, since this is an inviscid flow.

The flow accelerates as it flows around the front half of the Rankine body and decelerates around the rear half. This is evident from the streamlines shown in Fig. 5.9. Consequently, the pressure on the surface is the highest at the stagnation points A and B and lowest at the crown on the top and bottom surface. Since the flow is inviscid, the only force that acts on the body is pressure force. Since the flow field is symmetric about the vertical centerline, the force acting on the front half from left to right is exactly cancelled by the force acting on the rear half from right to left. Similarly, the symmetry of the flow field about the horizontal centerline of the body, results in the net force in the vertical direction being zero[‡].

The power of the superposition technique lies in the fact that new flows can be added (or removed) from existing solutions to generate new solutions. For example, removal of the sink from the above solution, results in the flow field shown in Fig. 5.10. This can be recognized as the flow around a Rankine half-body. On the other hand, removal of the source results in the flow field shown in Fig. 5.11.

Furthermore, variation of parameters such as the strength of the source, sink, distance between them, their relative orientation and the freestream

---

[†]This can be done without any impact or change in the flow field only when the fluid is inviscid. In real flows, such a replacement is not allowed, owing to the presence of viscous effects adjacent to such a surface.

[‡]In flows such as this, where an object is immersed in a stream, a net force on the object that acts horizontally from left to right *i.e. along the freestream direction* is usually called *drag*. A net force in the vertically upward direction is called *lift*.

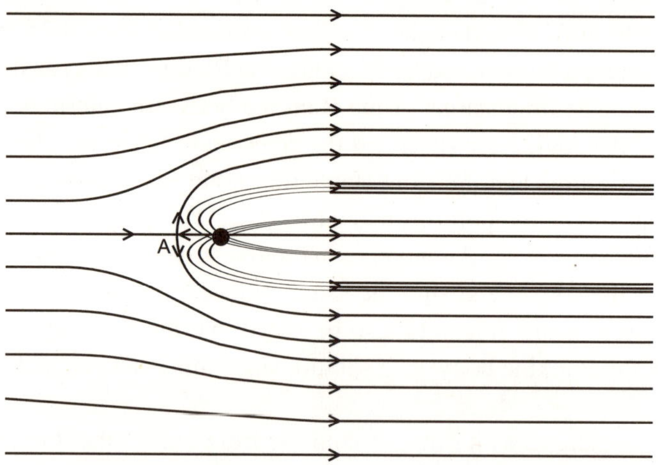

Fig. 5.10: Superposition of an uniform flow and a source

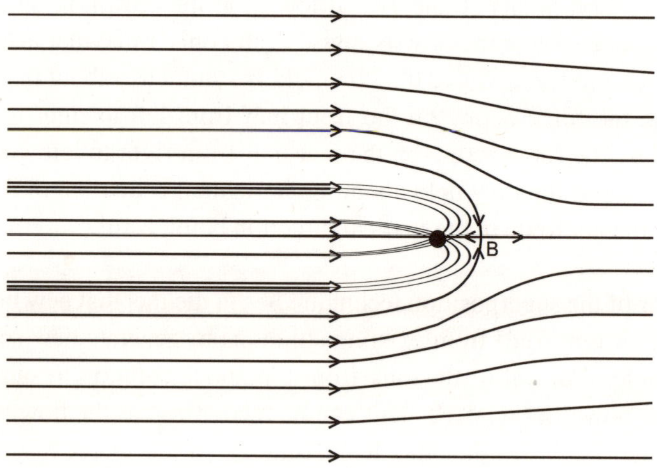

Fig. 5.11: Superposition of an uniform flow and a sink

velocity allows an almost infinite number of combinations (and solutions) to be realized.

Yet another solution that can be generated from the source-sink combination is the flow due to a doublet. This is accomplished by allowing the source and sink to move towards each other. Of course, they tend to partially cancel each other as they come closer and will identically

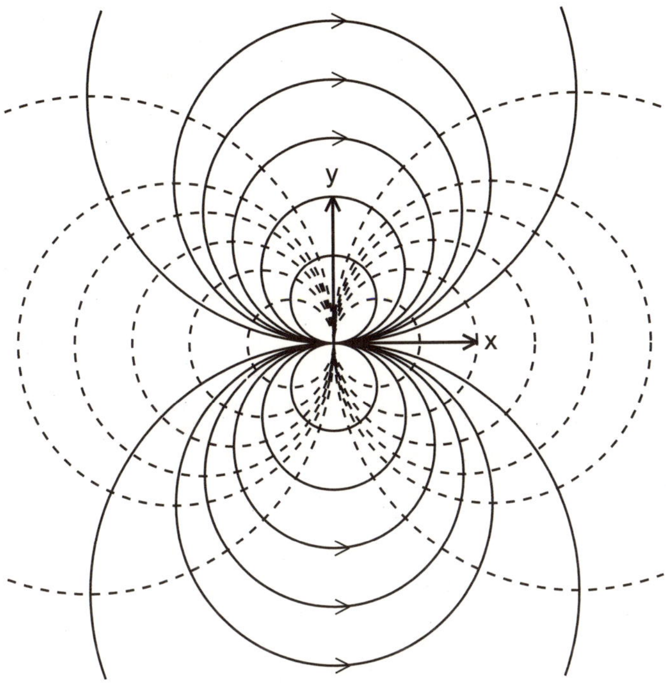

Fig. 5.12: Flow due to a doublet

cancel themselves when they are on top of each other. This can be avoided by increasing their strength in inverse proportion to the distance between them, *i.e.,* by keeping $2aq = constant = \Lambda$. The quantity Λ is called the strength of the doublet. The resulting flow field is that due to a doublet. By letting $q \to \infty$ and $a \to 0$ in Eqn. 5.24, the expression for the streamfunction for flow due to a doublet can be obtained. Thus

$$\psi = -\lim_{\substack{q \to \infty \\ a \to 0}} \left[q \tan^{-1}\left(\frac{2ay}{x^2 + y^2 - a^2} \right) \right] .$$

Since $\tan^{-1}\theta \approx \theta$ for small values of θ, the above expression can be written as

$$\psi = -\frac{\Lambda y}{x^2 + y^2} . \tag{5.27}$$

If we rearrange this equation and complete the squares, we get

$$x^2 + \left(y^2 + \frac{\Lambda}{2\psi}\right)^2 = \left(\frac{\Lambda}{2\psi}\right)^2 .$$

This equation describes, for different values of ψ, two families of circles with centers along the y-axis that are tangential to the x-axis at the origin, as sketched in Fig. 5.12. Equi-potential lines are shown as dashed lines in this figure.

Superposition of a uniform flow at zero angle of attack and a doublet

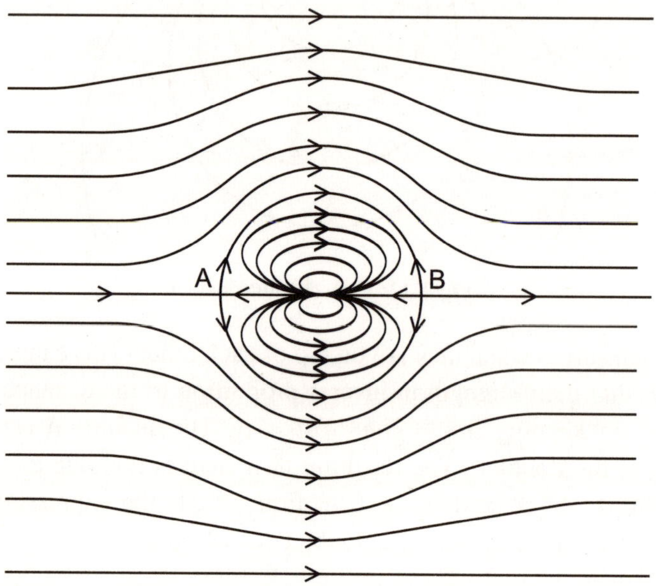

Fig. 5.13: Superposition of a uniform flow at zero angle of attack and a doublet

centered at the origin results in the flow field shown in Fig. 5.13. Similar to the case of the Rankine body, points A and B in this figure are stagnation points. The streamfunction for this flowfield is given as the sum of Eqns. 5.19 and 5.27,

$$\psi = U_\infty y - \frac{\Lambda y}{x^2 + y^2} . \tag{5.28}$$

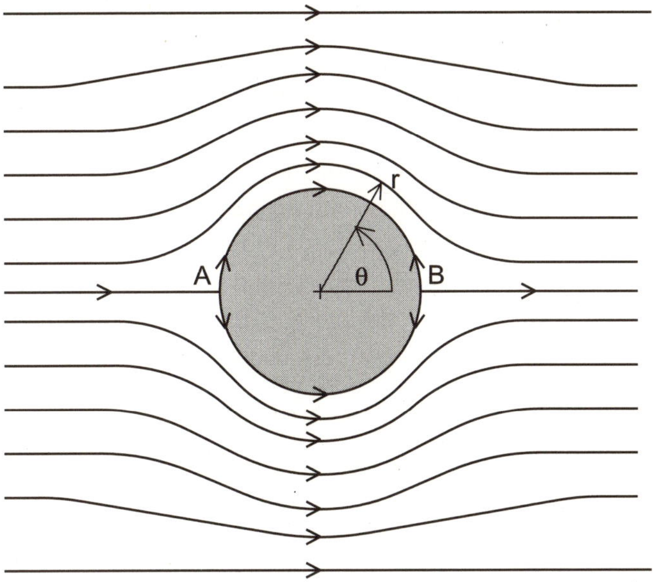

Fig. 5.14: Flow around a circular cylinder at zero angle of attack and without any circulation

If we switch to cylindrical polar coordinates, then

$$\psi = U_\infty r \sin\theta - \frac{\Lambda \sin\theta}{r} = U_\infty r \sin\theta \left(1 - \frac{a^2}{r^2}\right) , \tag{5.29}$$

where, we have used the fact that $y = r\sin\theta$ and $x^2 + y^2 = r^2$. Also, $a^2 = \Lambda/U_\infty$. The circular streamline ($\psi = 0$) passing through A and B completely encloses the flow inside it. Hence, if we replace it with an impermeable surface, we get the solution corresponding to the external flow around a circular cylinder as shown in Fig. 5.14.

Upon using Eqn. 2.5, we get

$$u_r = U_\infty \cos\theta \left(1 - \frac{a^2}{r^2}\right) \text{ and}$$

$$u_\theta = -U_\infty \sin\theta \left(1 + \frac{a^2}{r^2}\right) .$$

Since the normal component of velocity on the surface of the cylinder is zero, it is easy to see from the above expression that the radius of the cylinder is equal to a. The radius of the cylinder is thus seen to be determined by the magnitudes of the strength of the doublet and the freestream velocity. Since both u_r and u_θ are zero at the stagnation points, it follows then from the above expression that, $r = a$ and $\theta = 0, \pi$ are the locations of the stagnation points B and A respectively. It is easy to see from Eqn. 5.28 that, $\psi = 0$ represents the stagnation streamline. The pressure at any point in the flow field can be obtained from Bernoulli's equation. Thus

$$p = p_\infty + \frac{\rho U_\infty^2}{2} - \frac{\rho\left(u_r^2 + u_\theta^2\right)}{2} .$$

Of special interest is the pressure distribution on the surface of the cylinder, which can be obtained from the above expression by setting $r = a$. Thus,

$$p_s = p_\infty + \frac{\rho U_\infty^2}{2} - 2\rho U_\infty^2 \sin^2\theta = p_0 - 2\rho U_\infty^2 \sin^2\theta \,, \qquad (5.30)$$

where $p_0 = p_A = p_B = p_\infty + \rho U_\infty^2/2$ is the stagnation pressure and the subscript s denotes the surface. It is easy to see that the pressure reaches a minimum at $\theta = \pi/2$ and $3\pi/2$. Owing to the symmetry of the flow field about the horizontal and the vertical centerline of the cylinder, the net force exerted on the cylinder is zero. This result, which is contrary to the well known fact that a circular cylinder placed in a stream always experiences a drag, is known as the d'Alembert's paradox. It came to be known afterwards that the drag is due to the viscous force, which, of course, has been neglected in the development above.

If we superpose a clockwise vortex of strength Γ at the center of the cylinder, the resulting flow field looks as shown in Fig. 5.15. It is easy to infer from this figure that this is the flow field around a circular cylinder that is spinning about its own axis in the clockwise direction. The streamfunction for this flow field is obtained easily from Eqns. 5.31 and

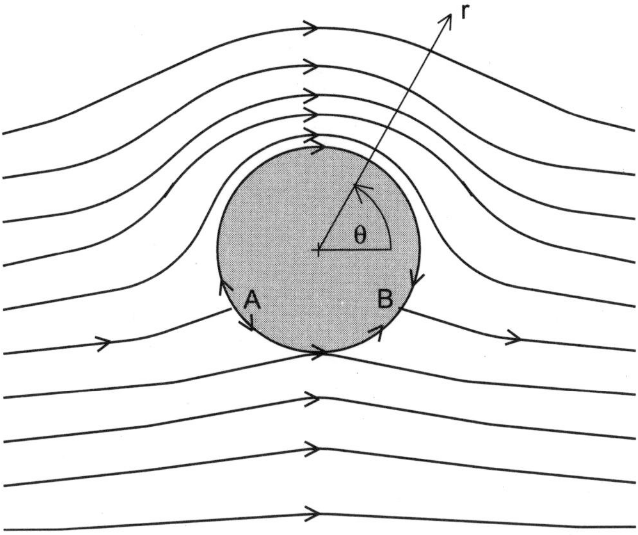

Fig. 5.15: Flow around a circular cylinder at zero angle of attack and with circulation $\Gamma < 4\pi U_\infty a$

5.20 as

$$\psi = U_\infty r \sin\theta \left(1 - \frac{a^2}{r^2}\right) + \frac{\Gamma}{2\pi} \ln r \, .$$

It is convenient to have the streamline corresponding to the surface of the cylinder be $\psi = 0$ as before. This can be accomplished by adding a constant to the right hand side of the above equation. Thus

$$\psi = U_\infty r \sin\theta \left(1 - \frac{a^2}{r^2}\right) + \frac{\Gamma}{2\pi} \ln\left(\frac{r}{a}\right) . \tag{5.31}$$

The radial and tangential velocity components for this case are given as

$$u_r = U_\infty \cos\theta \left(1 - \frac{a^2}{r^2}\right) \text{ and}$$

$$u_\theta = -U_\infty \sin\theta \left(1 + \frac{a^2}{r^2}\right) - \frac{\Gamma}{2\pi r} .$$

On the surface of the cylinder ($r = a$), $u_r = 0$ and $u_\theta = -2U_\infty \sin\theta - \Gamma/(2\pi a)$. The locations of the stagnation point(s) can be obtained by setting $u_\theta = 0$, which gives

$$\sin\theta_{A,B} = -\frac{\Gamma}{4\pi U_\infty}. \tag{5.32}$$

Fig. 5.16: Flow around a circular cylinder at zero angle of attack and with circulation $\Gamma = 4\pi U_\infty a$

For $\Gamma = 0$, there are two stagnation points and they are located at $\theta = 0$ and $\theta = \pi$, as shown in Fig. 5.14. For $0 < \Gamma < 4\pi U_\infty$, there are still two stagnation points, one in the third quadrant and one in the fourth quadrant, as shown in Fig. 5.15. As the circulation increases from 0 to $4\pi U_\infty$, the front and the rear stagnation points move towards each other. When $\Gamma = 4\pi U_\infty$, the two stagnation points merge into a single stagnation point at $\theta = 3\pi/2$ as shown in Fig. 5.16.

However, Eqn. 5.32 shows clearly that it is not possible to have stagnation point(s) on the surface of the cylinder for $\Gamma > 4\pi U_\infty a$. In this case, the

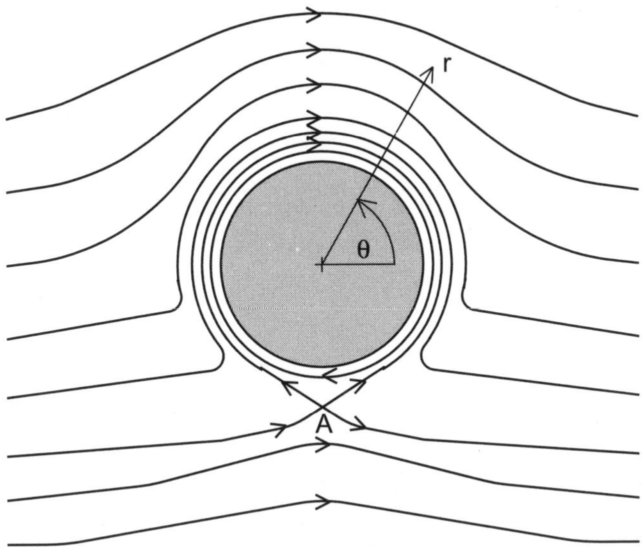

Fig. 5.17: Flow around a circular cylinder at zero angle of attack and with circulation $\Gamma > 4\pi U_\infty a$

stagnation point occurs in the flow field as shown in Fig. 5.17.

It is clear from Figs. 5.14-5.17 that the flow field loses its symmetry about the horizontal centerline for any non-zero value for the circulation. However, the flow field retains its symmetry about the vertical centerline. Consequently, the drag force on the cylinder is still zero but there is non-zero lift force on the cylinder. We now proceed to calculate this. If we apply Bernoulli's equation between a point in the freestream and on the cylinder, we get

$$p_\infty + \frac{\rho U_\infty^2}{2} = p_s + \frac{1}{2}\rho\left(2U_\infty \sin\theta + \frac{\Gamma}{2\pi a}\right)^2 .$$

This can be rearranged to give

$$p_s - p_\infty = \frac{\rho U_\infty^2}{2}\left(1 - 4\sin^2\theta - \frac{\Gamma \sin\theta}{2\pi a U_\infty} - \frac{\Gamma^2}{(2\pi a U_\infty)^2}\right) . \tag{5.33}$$

The normal force on a small element of the cylinder is $pd\mathcal{A}_n = (p_s -$

$p_\infty) a d\theta$. The net drag force on the cylinder, $\mathfrak{D}$, which is the horizontal component of the pressure force, can be calculated using

$$\mathfrak{D} = \int_0^{2\pi} (p_s - p_\infty) a \cos\theta \, d\theta \,.$$

Since $\cos\theta$ is an even function of θ, the above integral is equal to zero, as already surmised on physical grounds. The vertical component of the pressure force is the lift $\mathfrak{L}$, and this is given by

$$\mathfrak{L} = \int_0^{2\pi} (p_s - p_\infty) a \sin\theta \, d\theta \,.$$

Surprisingly, this turns out to be a very simple expression, *viz.*, $\mathfrak{L} = \rho U_\infty \Gamma$. Thus, there is a positive lift force on the cylinder†. This result does agree qualitatively with the well known *Magnus effect* that a spinning cylinder immersed in a freestream develops a lift force depending upon the direction of the spin relative to the freestream direction. The value for the lift obtained from the inviscid theory is much higher than experimentally measured values, however.

5.2.2.1 *Method of images*

All the solutions derived so far describe flows that are infinite in extent. The effect of confining wall (or walls) can be incorporated into these solutions by using a technique known as the method of images, which is also a superposition technique. The technique is demonstrated next for obtaining the flow field corresponding to a source located above a wall.

Let us assume that the source in Fig. 5.6 is placed at a height h above an infinitely long wall located on the x-axis. Surprisingly, the resulting flow field can be obtained quite easily by superposing the source flow field in Fig. 5.6 with that due to a source located directly below at $y = -h$. This is illustrated in Fig. 5.18.

In Fig. 5.18, it is easy to see that point A is a stagnation point.

†The lift force is negative *i.e.*, a net downward force, for a cylinder that spins in the clockwise direction.

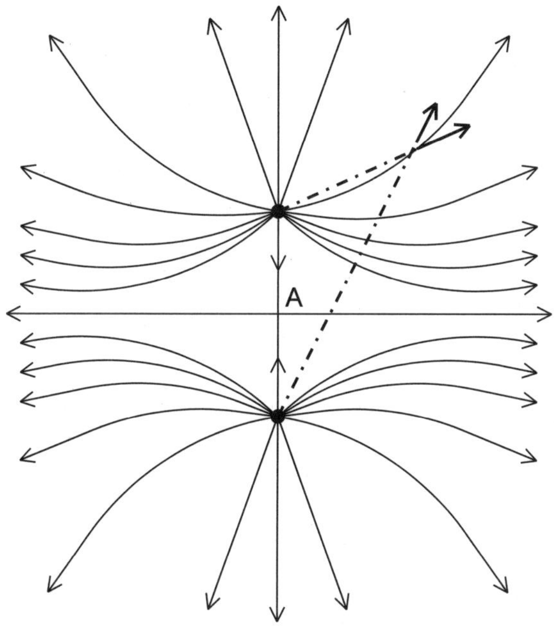

Fig. 5.18: Superposition of two sources of equal strength located one above the other

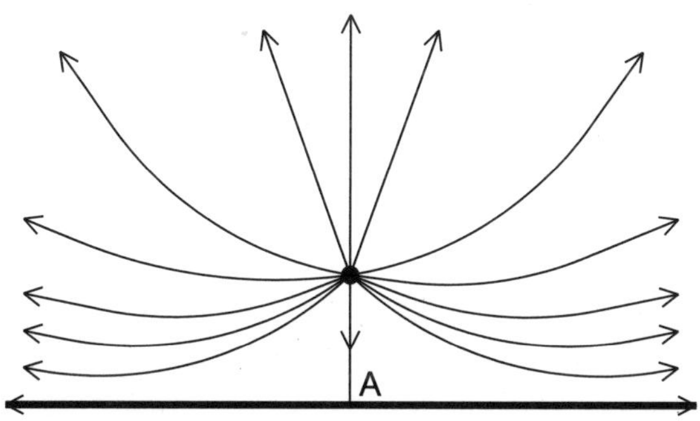

Fig. 5.19: Flow due to a source located above a wall

The horizontal streamline that passes through A, completely divides the flow above and below. Hence, this streamline can be replaced by an impermeable surface. If we do this and focus our attention on the flow field above the x-axis, it is clear that this represents the flow due to a source located above a wall (Fig. 5.19). The pressure is the highest at

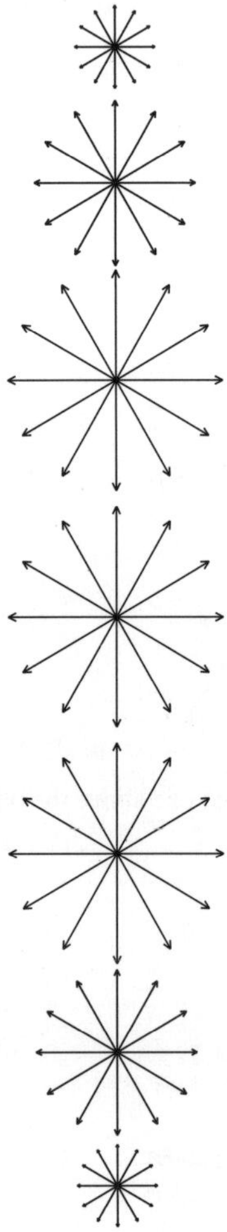

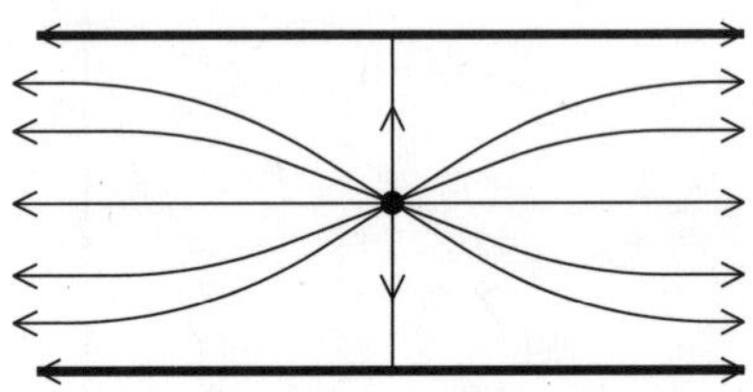

Fig. 5.20: Images used to generate the flow due to a source located in a channel (left) and the superposition of the flow due to the images (right)

the stagnation point A and falls off gradually on either side of it. Hence there is a net downward force on the surface.

In general, the effect due to the presence of a plane wall can be determined by creating a reflection (or mirror image) of the flow field about this plane and then superposing the two flow fields. In this manner, the effect of a wall on any of the flow fields discussed earlier, namely, flow over a Rankine body or flow over a circular cylinder, can be studied. We leave this as an exercise to the reader.

In principle, this idea can be extended to study the flow inside a channel (bounded by two walls). For instance, in the above example if we create a source at $y = 3h$ in addition to the one at $y = -h$, and superpose the resulting flow fields, we get what is seemingly the flow field due to a source placed in a channel of height $2h$. However, now, due to the fact the two reflecting surfaces (at $y = 0$ and $2h$) face each other, an infinite number of unwanted reflected images of diminishing strength are created†. These images would have to be cancelled by creating an equal number of images appropriately. This is illustrated in Fig. 5.20.

5.2.3 *Conformal mapping*

Conformal mapping is a powerful analytical technique to generate new solutions in a transformed coordinate system by transforming solutions from the Cartesian and cylindrical polar coordinate systems. The Joukowski transformation, for example, transforms the circle on a plane to a family of airfoil shapes depending upon the location of the center of the circle relative to the origin. Hence, the solution discussed above for the flow around a circular cylinder can be transformed into the flow around an airfoil. The airfoils can be of finite thickness or zero thickness, finite camber or zero camber and so on, depending upon the parameters of the transformation. However, it is not enough if we are merely able to transform shapes from one into another. In order for the transformed flow fields to be meaningful flow fields, the transformations should be conformal *i.e.,* they must satisfy certain mathematical requirements. The details are well beyond the scope of this book and we give only the gist here.

Conformal transformations,

- preserve angles
- preserve the strengths of sources, sinks and vortices

†Similar to what happens in a hair dressing salon when one sits between two mirrors that face each other.

- scale velocities

The first point ensures that, for example, flow that goes around an object (say, cylinder) in the original coordinate system, goes around the transformed object (say, airfoil) in the transformed coordinate system also and not *through* it. The Joukowski transformation mentioned above is a conformal transformation and is widely used. The Schwarz-Christoffel transformation is also a powerful conformal transformation.

5.2.4 *Flow in a sector*

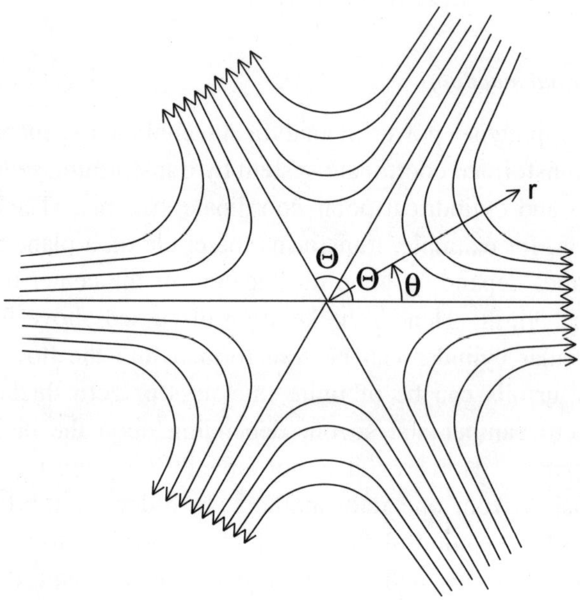

Fig. 5.21: Contours of the function $U_\infty r^n \sin(n\theta)$, for $n = 3$. Note that $\Theta = \pi/n$.

The solutions that we have discussed so far have been largely obtained using the principle of superposition. In this section, some solutions that can be obtained without using this principle are presented.

The function $\sin(n\theta)$ is periodic with period $2\pi/n$, and changes sign every $\Theta = \pi/n$. The function $\psi = r^n \sin(n\theta)$, where r and θ are the radial

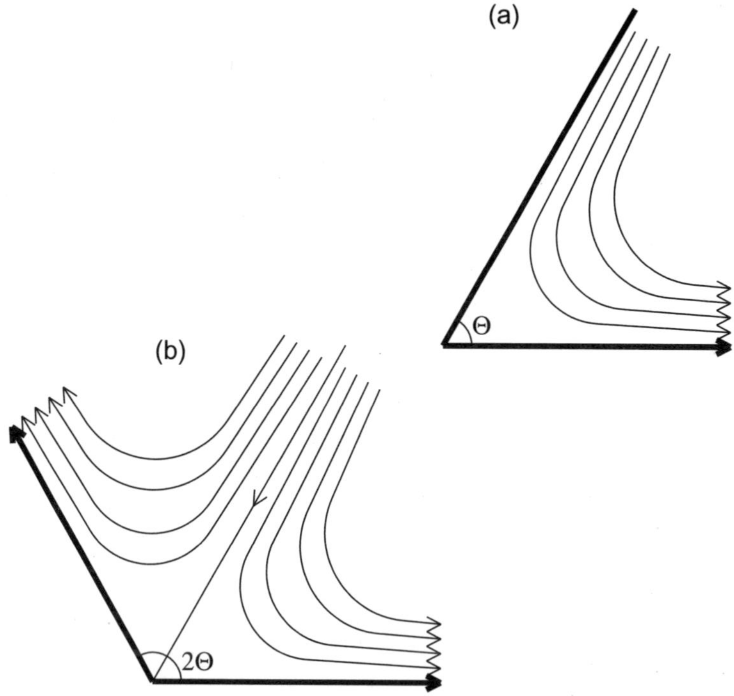

Fig. 5.22: (a) Flow in a 60° sector and (b) Stagnation flow towards a 120° sector

and azimuthal coordinates respectively, also exhibits the same behavior, and the contours of this function for $n = 3$ are shown in Fig. 5.21. The velocity components can be obtained using Eqn. 2.5 as

$$u_r = nr^{n-1}\cos(n\theta) \quad \text{and} \quad u_\theta = -nr^{n-1}\sin(n\theta)\,. \tag{5.34}$$

The radial velocity $u_r = nr^{n-1}$ and $u_\theta = 0$ along $\theta = 0, \Theta, 2\Theta, 3\Theta, \cdots$. Hence, the streamlines at $\theta = 2\Theta, 4\Theta, \cdots$ are radial lines across which the flow field is periodic, and the streamlines at $\theta = 0, \Theta, 3\Theta, \cdots$ are also radial and the tangential component of velocity changes sign across them. In a manner similar to what was done earlier, these streamlines can be replaced with impermeable surfaces and the resulting flow fields can be interpreted as the flow in a 60° corner and the plane stagnation flow towards a 120° corner. These are shown in Fig. 5.22. Flow fields corresponding to $n = 2$, namely, flow in a 90° corner and plane stagnation flow towards a

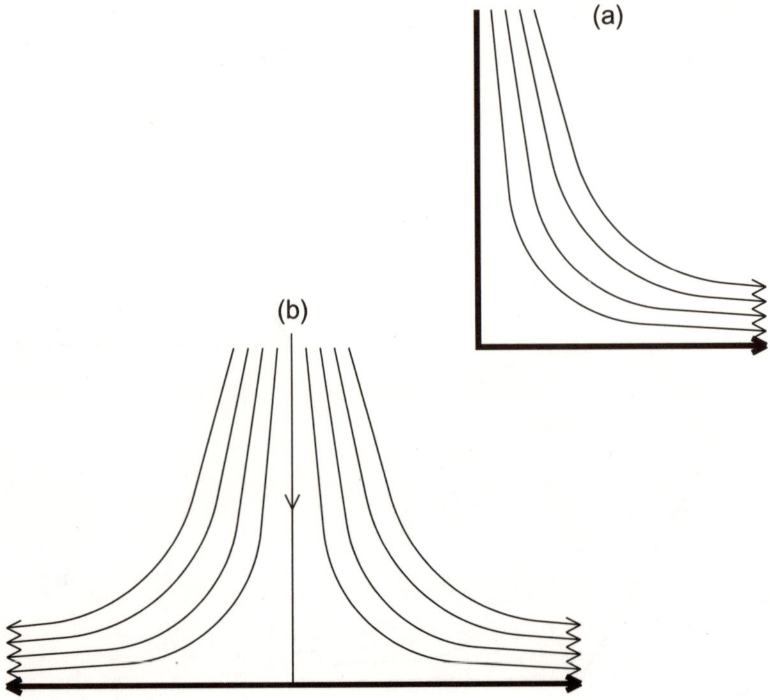

Fig. 5.23: (a) Flow in a 90° sector and (b) Plane stagnation flow towards a wall

wall are shown in Fig. 5.23†.

The potential flow solutions for the stagnation flow shown in Figs. 5.22 and 5.23 are very good solutions since viscous effects are confined to thin boundary layers adjacent to the wall. In fact, the matching of the potential flow solution for the plane stagnation flow towards a wall with the boundary layer solution will be demonstrated in the next chapter. On the other hand, the usefulness of the potential flow solutions in the corners, namely, Figs. 5.22(a) and 5.23(a) is somewhat limited. This is due to the fact that viscous effects are important not only near the walls but in other regions as well, owing to the proximity of the walls. As the flow approaches the corner, it separates completely from both the walls and forms a recirculation region that occupies the entire cross-section of the

†Fractional values for n are also allowed. Interested readers may try to sketch the flow field for n=2/3, 3/2 and 8/7.

sector. This flow too, in turn, separates from the wall further near the corner and so on. This results in an infinite number of corner vortices of diminishing size called Moffatt vortices†.

The 2D potential flow discussed in this chapter form a very small subset of the entire set of possible solutions. Axi-symmtetric solutions, which are also relatively easily obtained have not been discussed here. Owing to the ubiquity of the potential equation 5.18, and its axi-symmteric counterpart, a great amount of analytical and numerical solutions are available in the literature. In flows, where the viscous effects are confined to small regions, such as the flow over objects and stagnation flows, the potential flow solutions serve as good outer solutions which could then be patched up with boundary layer solutions. The sophistication of potential flow theory has grown over the years to the extent that today even massively separated flows can be *modeled* quite well. The interested reader is urged to consult advanced text books on inviscid flow theory for more information.

†see the book *An Album of Fluid Motion* by M. Van Dyke, Parabolic Press, 1982 for a nice visualization.

Exercises

(1) A large tank contains three layers of different liquids as shown in the figure. By applying Bernoulli's equation along a streamline starting from the free surface, determine the discharge velocity. The flow may be assumed to be steady and inviscid.

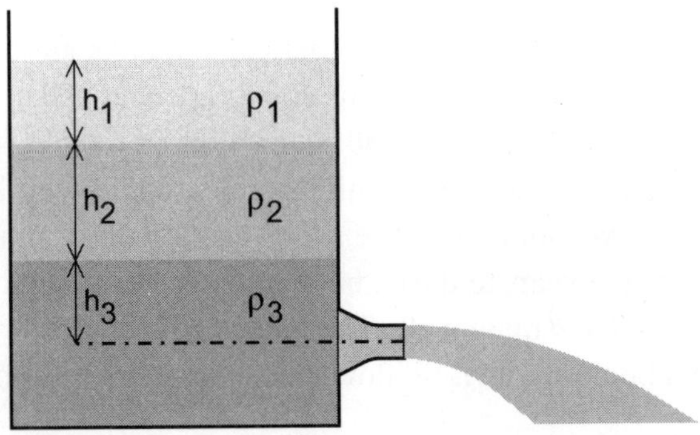

(2) A pipe of length L is connected to a large tank containing a liquid as shown in the figure. The pipe is initially closed. Determine the velocity of the fluid as a function of time at the exit of the pipe after it is opened. Assume the fluid to be inviscid and ignore any entrance effects near the pipe entrance.

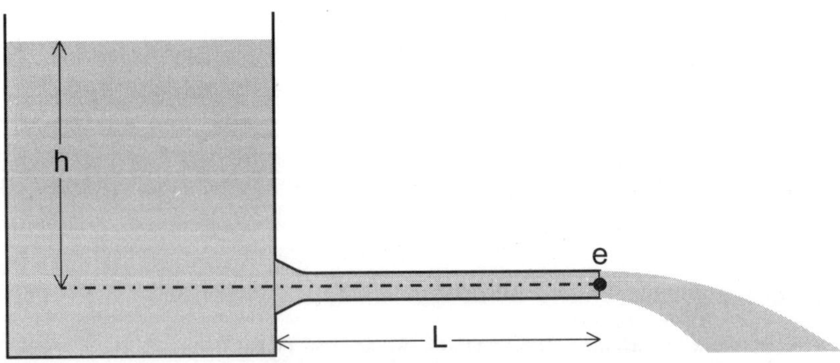

(3) A U-tube intially contains a liquid to height h in each limb. The liquid in one limb is displaced slightly and released. Assuming the fluid to be inviscid, determine the period of oscillation of the liquid column by applying the unsteady Bernoulli equation. The length of the curved portion of the U-tube may be neglected.

(4) Consider the flow over a hut, the cross-section of which is modelled as a semi-circle. The pressure inside the hut is p_i. Determine the net force on the hut and establish that it acts vertically upwards. By making a small hole on the roof of the hut, the pressure inside can be made equal to the surface pressure at the location where the hole is made. Determine the angle at which the hole should be made, so that the net force on the hut is zero. [$\mathcal{L} = 2a(P_\infty - p_i) - \frac{5}{3}\rho_\infty a U_\infty^2, \theta = 54.735°$]

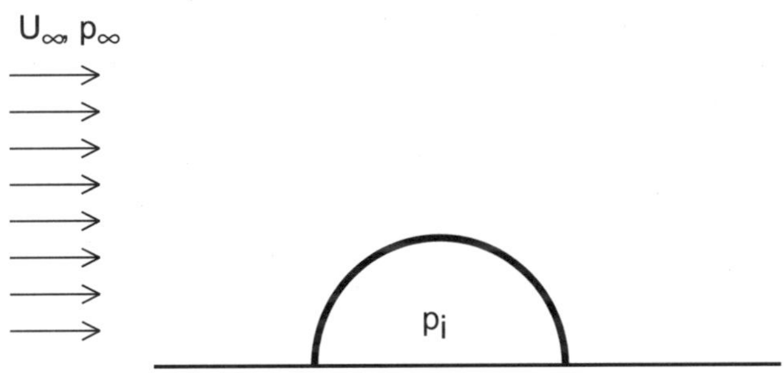

(5) Obtain expressions for the velocity components for the inviscid flow over a wedge with an included angle of $\pi/3$.

(6) A cylinder of radius a is located a distance $3a$ above a wall as shown in the figure. Assuming the flow to be steady and inviscid, determine the velocity and pressure distribution on the surface of the cylinder. Establish by symmetry arguments whether the lift and drag forces on the cylinder are zero or not. Determine the non-zero lift and/or drag force on the cylinder. Also determine the pressure and velocity at the locations shown in the figure and compare with the corresponding values when the wall is not present.

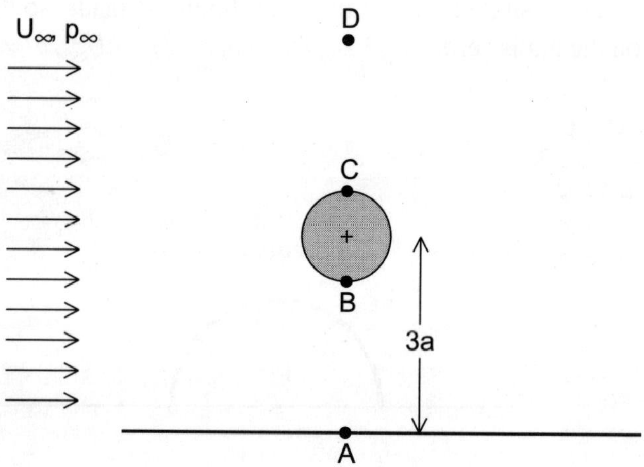

U∞, p∞
D
C
B
3a
A

Chapter 6

Laminar Boundary Layer Theory

In this chapter, the boundary layer equations that arise from the simplification of the incompressible Navier-Stokes equations under the boundary layer approximation are derived. Solutions to these equations using differential and integral techniques for different types of outer (potential) flows are presented. Separation of the boundary layer and its consequences are discussed at the end.

6.1 Derivation of the boundary layer equations

We start with the continuity equation 3.11 and the incompressible Navier-Stokes equations 4.1. Thus

$$\frac{\partial u}{\partial x} + \frac{\partial v}{\partial y} = 0\,, \tag{6.1}$$

$$u\,\frac{\partial u}{\partial x} + v\,\frac{\partial u}{\partial y} = -\frac{1}{\rho}\frac{\partial p}{\partial x} + \nu\left(\frac{\partial^2 u}{\partial x^2} + \frac{\partial^2 u}{\partial y^2}\right)\,, \tag{6.2}$$

and

$$u\,\frac{\partial v}{\partial x} + v\,\frac{\partial v}{\partial y} = -\frac{1}{\rho}\frac{\partial p}{\partial y} + \nu\left(\frac{\partial^2 v}{\partial x^2} + \frac{\partial^2 v}{\partial y^2}\right)\,, \tag{6.3}$$

where, we have assumed the flow to be steady and two dimensional and neglected gravity also, without any loss of generality.

We now wish to enquire whether these equations have to be used as they are, or can be simplified, in the inner(boundary) layer next to a no-slip surface. This can be ascertained using an order of magnitude analysis of Eqns. 6.1-6.3, similar to what is discussed in Section 4.3. Of course, the primary intent is to apply the no-slip boundary condition, in contrast to the flows discussed in the previous chapter. We assume that the no-slip surface is stationary, flat and lies on the x-axis. The thickness of the inner layer is taken to be δ and the x-velocity outside this layer is taken to be U_∞ and the y-velocity is zero (Fig. 6.1). Note that, U_∞ itself would be known from the potential flow solution. We assume that the characteristic dimension along the x-coordinate is L (this can be the length of the no-slip surface, for example). The most important requirement is that $L \gg \delta$. Any change in a flow variable takes place across a length scale L if it is along the x-direction and δ if it is along the y-direction. Since the x-component of velocity is zero on the no-slip surface and U_∞ outside the boundary layer (freestream), the maximum change in the x-component of velocity is U_∞. It only remains to determine whether or not v is zero. This is easy (and insightful) to establish using a physical argument as follows.

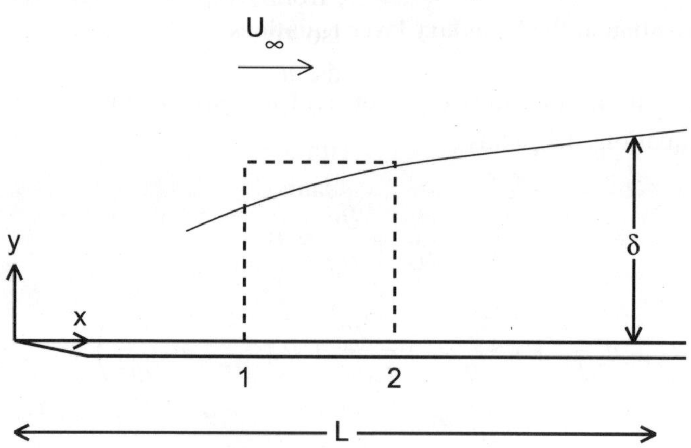

Fig. 6.1: Illustration of the inner layer near a no-slip surface

It seems at first sight that the y-velocity in the boundary layer will likely be zero, since it is zero on the no-slip surface as well as outside far away in the freestream. As illustrated in Fig. 6.1, the thickness of the boundary layer increases (in contrast to the example in Section 4.3), since the presence of the no-slip surface decelerates the fluid in the boundary layer continuously. In other words, the fluid that has already been decelerated to some extent at station 1 undergoes further deceleration by the time it reaches station 2 and hence the thickness of the boundary layer at station 2 is more than it is at station 1. Now consider the control volume shown with dashed line in Fig. 6.1. Owing to the deceleration of the fluid between stations 1 and 2, the mass that enters through the left face of the control volume is more than the mass that leaves through the right face. Since no mass enters or leaves through the bottom face, for conservation of mass within the control volume, it is clear that this excess mass has to leave through the top face. This means that the normal component of velocity on the top surface, which, in this case is the y-component, has to be non-zero and *positive*. It is thus reasonable to surmise that the y-component of velocity at an $x = constant$ station is zero on the no-slip surface, increases thereafter and reaches a maximum value at the edge of the boundary layer[†].

This can be established, alternatively, from Eqn. 6.1. If we assume v to be zero everywhere, then $\partial v/\partial y$ is also zero. It then follows from Eqn. 6.1 that $\partial u/\partial x$ is zero. In other words, u is a function of y alone. This is not tenable, since the continuous deceleration of the fluid by the no-slip surface results in u decreasing with x along any $y = constant$ line inside the boundary layer. Hence, v cannot be zero inside the boundary layer. Furthermore, since u decreases with x inside the boundary layer, $\partial u/\partial x$ is negative. Hence, $\partial v/\partial y$ is positive from Eqn. 6.1. This implies that v increases inside the boundary layer from zero on the no-slip surface, to a maximum at the edge of the boundary layer as we move along an $x = constant$ line, as mentioned before.

We are now ready to proceed with the scaling analysis. The sizes of the two terms in Eqn. 6.1 are

$$\frac{U_\infty}{L} \qquad \frac{V}{\delta},$$

where, V is an estimate (as yet unknown) of the magnitude of the y-

[†]Note that v cannot change outside the boundary layer since the flow is no longer decelerated by the no-slip surface.

component of velocity inside the boundary layer. Both the terms must be of the same size since, neither $\partial u/\partial x$, nor $\partial v/\partial y$ is zero†. Hence

$$V \sim \frac{U_\infty \delta}{L}.$$

Since $\delta \ll L$, the y-component of velocity in the boundary layer while non-zero, is much smaller than U_∞. In fact, the above expression makes it clear that y-component of the velocity inside the boundary layer is an order of magnitude (δ) less than the streamwise (x-) component of velocity. The sizes of the terms in Eqn. 6.2 are

$$\frac{U_\infty^2}{L} \quad \frac{VU_\infty}{\delta} \quad \frac{P}{\rho L} \quad \frac{\nu U_\infty}{L^2} \quad \frac{\nu U_\infty}{\delta^2},$$

where P is an estimate (as yet unknown) of the pressure in the boundary layer. If we substitute for V, we get

$$\frac{U_\infty^2}{L} \quad \frac{U_\infty^2}{L} \quad \frac{P}{\rho L} \quad \frac{\nu U_\infty}{L^2} \quad \frac{\nu U_\infty}{\delta^2}.$$

If we compare the two viscous terms, it is clear that the term with the δ^2 in the denominator is much larger than the other one, since $\delta \ll L$. Hence the term $\nu \partial^2 u/\partial x^2$ can be dropped. The sizes of the remaining terms are

$$\frac{U_\infty^2}{L} \quad \frac{U_\infty^2}{L} \quad \frac{P}{\rho L} \quad \frac{\nu U_\infty}{\delta^2}.$$

The two convective terms are of the same size and we definitely need to retain the remaining viscous term. So, we equate the size of the convective and the viscous term. This requires that

$$\frac{\delta}{L} \sim \sqrt{\frac{\nu}{U_\infty L}}. \tag{6.4}$$

†The two terms must be opposite in sign since their sum is zero.

The size of the pressure gradient term is, as of yet, undetermined. Since all the other terms are of magnitude U_∞^2/L, the pressure gradient term also has to be the same size if it has to be retained. Thus

$$\frac{P}{\rho L} \sim \frac{U_\infty^2}{L} \qquad \Rightarrow P \sim \rho U_\infty^2 \,.$$

The sizes of the terms in Eqn. 6.3 are

$$\frac{U_\infty V}{L} \qquad \frac{V^2}{\delta} \qquad \frac{P}{\rho\delta} \qquad \frac{\nu V}{L^2} \qquad \frac{\nu V}{\delta^2} \,.$$

Upon substituting for V, we get

$$\frac{U_\infty^2 \delta}{L^2} \qquad \frac{U_\infty^2 \delta}{L^2} \qquad \frac{P}{\rho\delta} \qquad \frac{\nu U_\infty \delta}{L^3} \qquad \frac{\nu U_\infty \delta}{\delta^2 L} \,.$$

Upon rearranging, we have,

$$\underline{\frac{U_\infty^2}{L}\frac{\delta^2}{L}} \qquad \underline{\frac{U_\infty^2}{L}\frac{\delta^2}{L}} \qquad \frac{P}{\rho} \qquad \underline{\frac{U_\infty^2}{L}\frac{\delta^2}{L}\frac{\delta^2}{L^2}} \qquad \underline{\frac{U_\infty^2}{L}\frac{\delta^2}{L}} \,,$$

where, we have set $\nu U_\infty/\delta^2 = U_\infty^2/L$. The convective terms and the viscous terms (shown underlined) are at least two orders of magnitude (δ^2) less than the pressure gradient term and hence can be neglected. We are thus left only with the pressure gradient term, and Eqn. 6.3 becomes

$$\frac{\partial p}{\partial y} = 0 \,. \tag{6.5}$$

This has certain important implications:

- The pressure gradient term has to be retained in the x-momentum equation, as otherwise, pressure will completely drop out of the governing equations.

- Since p does not change in the y-direction, the pressure inside the boundary layer is the same as that outside in the freestream, at the edge of the boundary layer *i.e.,* $p = p_\infty$. It must be recalled that p_∞ is available from the potential flow solution through the Bernoulli equation.

- The pressure inside the boundary layer can be a constant or a function of x only.

The inner (boundary) layer equations are thus, Eqns. 6.1, 6.5 and

$$u\,\frac{\partial u}{\partial x} + v\,\frac{\partial u}{\partial y} = -\frac{1}{\rho}\frac{dp_\infty}{dx} + \nu\frac{\partial^2 u}{\partial y^2}\,. \tag{6.6}$$

In writing the above equation, p has been set equal to p_∞ and the partial derivative has been replaced with an ordinary derivative, since p is a function of x only. The pressure can be completely eliminated by using the Bernoulli equation, since

$$\frac{U_\infty^2}{2} + \frac{p_\infty}{\rho} = constant$$

for a potential flow. Upon differentiating this, we get $dp_\infty/dx = -\rho U_\infty\, dU_\infty/dx$. Hence, Eqn. 6.6 can be written as

$$u\,\frac{\partial u}{\partial x} + v\,\frac{\partial u}{\partial y} = U_\infty\frac{dU_\infty}{dx} + \nu\frac{\partial^2 u}{\partial y^2}\,. \tag{6.7}$$

Thus, the boundary layer equations are Eqns. 6.1 and 6.7, subject to the boundary conditions,

$$u(x, y = 0) = 0\,, \qquad u(x, y \to \infty) \to U_\infty \tag{6.8}$$

and $v(x, y = 0) = 0$. Note that the second (matching) condition in Eqn. 6.8 is applied not at $y = \delta$, as one might expect. Although this point is discussed in Section 4.3, it is worth reiterating here. The scaling analysis suggests that the boundary layer thickness is only proportional - *not equal*, to $\sqrt{\nu/U_\infty L}$. Therefore, it is mathematically correct to apply the matching condition as the inner coordinate tends to ∞. Strictly speaking, the right hand side of the matching condition must say that it is to be evaluated as the outer coordinate tends to zero, although we have not indicated this explicitly in Eqn. 6.8. Once the solution to the boundary layer equations has been obtained, the thickness of the boundary layer (actually, the proportionality constant in Eqn. 6.4) can be determined by using a suitable criterion. The y-location at which $u = 0.99U_\infty$ is frequently taken as the edge of the boundary layer. Alternatively, the y-location at which $\partial u/\partial y$ becomes zero can be taken as the edge of the boundary layer. This is based on the physical argument that the shear stresses have to be zero in the inviscid outer layer.

Before we discuss solutions for the boundary layer equations, some observations regarding the equations are in order.

- Eqn. 6.1 is exact whereas Eqn. 6.7 is an approximation to Eqns. 6.2 and 6.3. Hence, the boundary layer equations are an approximation to the incompressible Navier-Stokes equations inside the boundary layer.

- $\partial u/\partial x$ is retained in Eqns. 6.1 and 6.7, while $\nu\, \partial^2 u/\partial x^2$ is neglected. Similarly, v and $\partial v/\partial y$ are retained in these equations, while all the terms containing v are neglected in Eqn. 6.3. The right hand side in Eqn. 6.5 is actually $O(\delta^2)$, but has been neglected since δ is very small. Solutions to the Navier-Stokes equations will show $\partial p/\partial y$ to be small but non-zero. The key assumption is that since $\delta \ll L$, terms containing derivatives in the x-direction are small compared to terms containing derivatives in the y-direction. If this assumption is violated†, then the boundary layer equations have to be modified.

- The two boundary layer equations, namely, Eqns. 6.1 and 6.7 can in principle be used to solve for the two unknowns u and v. However, there is no equation as yet for determining the remaining unknown quantity, namely, the boundary layer thickness, δ. Such an equation has to be generated by some other means.

---

†For example, near a separation point or in flows in which curvature effects are very strong so that streamwise derivatives are comparable in size to the cross-stream derivatives.

The above development is mathematical in nature as it has followed singular perturbation theory. However, the credit for hypothesizing the existence of a boundary layer next to solid surfaces on physical grounds and the subsequent development of the theory should go to Prandtl.

An important consequence of the imposition of the no slip boundary condition is, of course, the formation of the boundary layer. Yet another important consequence is the generation of vorticity in the boundary layer owing to the fact that the shear stresses are non-zero. Since we are dealing with a 2D flow, the only non-zero component of vorticity is ω_z and this is given as (Eqn. 2.8)

$$\omega_z = \frac{\partial v}{\partial x} - \frac{\partial u}{\partial y}.$$

An order of magnitude estimate shows that the two terms are

$$\frac{U_\infty \delta}{L^2} \qquad \frac{U_\infty}{\delta}$$

in size. Upon rearranging, we can see that the two terms are

$$\frac{U_\infty \delta^2}{L^2} \qquad U_\infty$$

in size. It is clear that the first term is two orders of magnitude less than the second one and can be neglected. Thus $\omega_z = -\dfrac{\partial u}{\partial y}$. Since $\partial u/\partial y$ is a maximum at the wall and decays to zero at the edge of the boundary layer and beyond, vorticity follows the same trend.

6.2 Boundary layer flow over a flat plate with zero pressure gradient

In this section, solution to the boundary layer equations for the flow over a flat plate with zero pressure gradient is obtained. The absence of the pressure gradient simplifies the equations to some extent. Flows with non-zero pressure gradients are considered in the next section.

If the pressure gradient is zero, then the pressure is constant and equal to p_∞ everywhere. It follows then from Bernoulli's equation that U_∞ is also a constant. Therefore, dU_∞/dx is zero and Eqn. 6.7 simplifies to

$$u\frac{\partial u}{\partial x}+v\frac{\partial u}{\partial y}=\nu\frac{\partial^2 u}{\partial y^2}\,. \tag{6.9}$$

It is important to understand that the zero pressure gradient flow is an idealization. With reference to the control volume shown in Fig. 6.1, it is easy to see that the x-momentum of the fluid leaving through the right face is less than that coming in through the left face. Of course, a small amount of x-momentum is carried away by the fluid leaving through the top face. Notwithstanding this, there is a net reduction in the momentum of the fluid as it passes through the control volume, which is exhibited as a drag force *on* the fluid in the negative x-direction†. The fluid thus loses momentum continuously as it flows over the surface. Unless there is a favorable pressure gradient to maintain the flow, it will stagnate very quickly. In the context of our discussion leading up to Eqn. 6.9, we assumed that the momentum extracted by the plate is so small, that the pressure gradient required to drive the flow is negligibly small‡.

6.2.1 *Differential analysis - Principle of similarity*

Boundary layer flow over a flat plate with zero pressure gradient is illustrated in Fig. 6.2. Note that U_∞ remains a constant. The velocity profiles across the boundary layer are shown at two different stations 1 and 2. It must be kept in mind that although these velocity vectors appear to be entirely horizontal, there is a small and positive y-velocity component. The thickness of the boundary layer at these two locations are also shown as δ_1 and δ_2. The same velocity profiles are shown on the left in Fig. 6.3 on a slightly enlarged scale for the sake of clarity. Let the velocity at $x = x_1$, $y = y_1$ be equal to the velocity at $x = x_2$, $y = y_2$ as shown in Fig. 6.3 on the left. It turns out that the velocity profiles are self-similar, that is, $y_1/\delta_1 = y_2/\delta_2$. In other words, instead of plotting y against u as shown

†From Newton's third law, the drag force on the plate is equal in magnitude but in the opposite direction *i.e.,* the positive x-direction. Hence the plate will be blown away unless it is firmly held in place through an external force.

‡Such a flow is extremely difficult to realize in practice. See the book *An Album of Fluid Motion* by M. Van Dyke, Parabolic Press, 1982 for a nice visualization of such a flow.

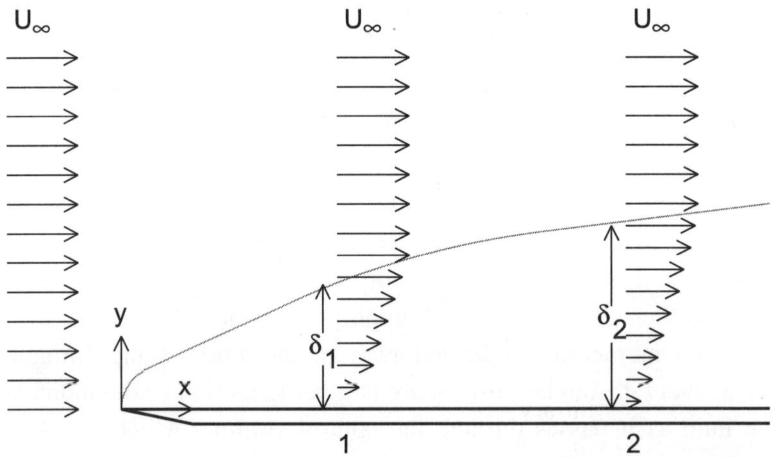

Fig. 6.2: Illustration of the boundary layer flow over a flat plate with zero pressure gradient

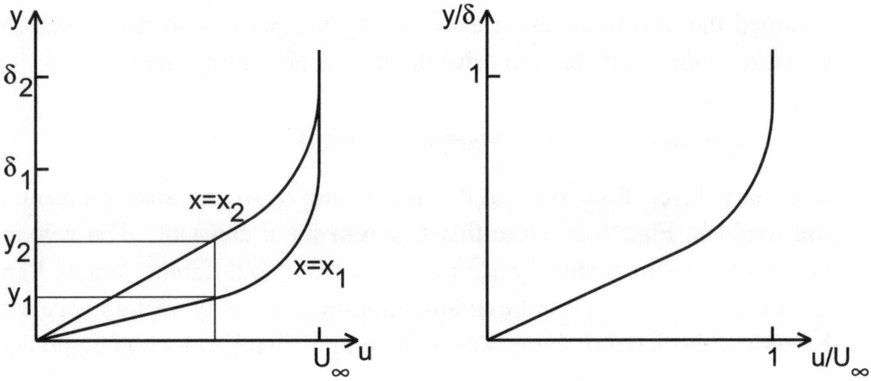

Fig. 6.3: Self similar profile of u/U_∞

in Fig. 6.3 on the left, if we plot y/δ against u/U_∞, then the velocity profiles at all values of x collapse on to a single curve as shown on the right in the same figure. This signifies that, although u is a function of two independent variables x and y, u/U_∞ is a function of one independent variable $\eta = y/\delta(x)$ only. Consequently, it is reasonable to expect that the partial differential equations Eqn. 6.1 in x and y and 6.9 can be written as ordinary differential equations in η.

At this point, the concept of self-similarity has been introduced as a hypothesis only. It remains to be established whether a self-similar solution to Eqns. 6.1 and 6.9 is admissible. This is taken up next.

We start by integrating $u = \dfrac{\partial \psi}{\partial y}$ from Eqn. 2.4,

$$\psi = \int_0^y u dy + k(x) \,,$$

where $k(x)$ is an arbitrary function in x. Let $u/U_\infty = h(\eta)$, where h is an unknown function. The above integral becomes

$$\psi = U_\infty \delta(x) \int_0^\eta h(\xi) d\xi + k(x) \,,$$

where ξ is a dummy integration variable and we have used the fact that $dy = \delta(x) d\eta$. Since the flat plate itself is a streamline (because no flow crosses it), we can conveniently set $\psi = 0$ at $y = 0$. Therefore, $k(x)$ in the above equation is equal to zero. Thus

$$\psi = U_\infty \delta(x) \int_0^\eta h(\xi) d\xi = U_\infty \delta(x) f(\eta) \,,$$

where we have conveniently replaced the integral with $f(\eta)$. Now,

$$\begin{aligned} u &= \frac{\partial \psi}{\partial y} = U_\infty \delta(x) \frac{\partial f}{\partial y} = U_\infty \delta(x) \frac{\partial \eta}{\partial y} \frac{df}{d\eta} = U_\infty \delta(x) \frac{1}{\delta(x)} f' \\ &= U_\infty f' \,, \end{aligned} \tag{6.10}$$

where $'$ denotes differentiation with respect to η. Similarly,

$$
\begin{aligned}
v = -\frac{\partial \psi}{\partial x} &= -U_\infty \left(f \frac{d\delta}{dx} + \delta(x) \frac{\partial f}{\partial x} \right) \\
&= -U_\infty \left(f \frac{d\delta}{dx} + \delta(x) \frac{\partial \eta}{\partial x} \frac{df}{d\eta} \right) \\
&= -U_\infty \left(f \frac{d\delta}{dx} - \delta(x) \frac{y}{\delta^2(x)} \frac{d\delta}{dx} f' \right) \\
&= -U_\infty \frac{d\delta}{dx} \left(f - \eta f' \right) .
\end{aligned} \tag{6.11}
$$

In deriving the above equation, we have used the fact that U_∞ is a constant and not a function of x. If we substitute the above expressions into Eqn. 6.9, we get

$$
U_\infty f' \frac{\partial}{\partial x}(U_\infty f') - U_\infty \frac{d\delta}{dx} \left(f - \eta f' \right) \frac{\partial}{\partial y}(U_\infty f') = \nu \frac{\partial^2}{\partial y^2}(U_\infty f') .
$$

If the partial derivatives with respect to x and y are converted into ordinary derivatives with respect to η, we get

$$
-U_\infty^2 f' f'' \frac{y}{\delta^2(x)} \frac{d\delta}{dx} - U_\infty^2 \left(f - \eta f' \right) \frac{d\delta}{dx} \frac{f''}{\delta(x)} = \nu U_\infty \frac{f'''}{\delta^2(x)} .
$$

This can be simplified as

$$
f''' + \underline{\frac{U_\infty \delta(x)}{\nu} \frac{d\delta}{dx}} f f'' = 0 .
$$

Since we expect the solution to be self-similar in η, the above equation can contain either constants or quantities that depend on η alone. Hence, the underlined term has to be a constant or a function of η. The latter possibility is ruled out, since both δ and $d\delta/dx$ are functions of x alone. Therefore, for a self-similar solution to exist, we demand that

$$
\frac{U_\infty \delta(x)}{\nu} \frac{d\delta}{dx} = c ,
$$

and

$$f''' + cff'' = 0\,,$$

where c is a constant. The power and elegance of the self-similarity principle is brought out by these two equations, since it provides an equation that governs the boundary layer thickness and also simplifies the original system of partial differential equation to a single, ordinary differential equation (albeit non-linear).

If we integrate the first of these equations, we get

$$\frac{1}{2}\delta^2 = \frac{c\nu}{U_\infty}\,x + d\,,$$

where, d is an arbitrary constant. Since $\delta = 0$ at $x = 0$, (See Fig. 6.2), $d = 0$. Therefore[†]

$$\delta = \sqrt{\frac{2c\nu x}{U_\infty}}\,.$$

This still leaves the constant c undetermined, which is not a serious issue, since the functional form has been determined. The thickness of the boundary layer, anyway, has to determined using one of the criteria mentioned earlier and this is possible only after the boundary layer equation has been solved. It is customary to take $c = 1/2$ and so

$$\delta = \sqrt{\frac{\nu x}{U_\infty}}\,, \tag{6.12}$$

and

$$f''' + \frac{1}{2}\,ff'' = 0\,. \tag{6.13}$$

This equation was first derived by Blasius, a student of Prandtl. Boundary conditions for this equation follow from Eqn. 6.8, *viz.*,

$$f(0) = f'(0) = 0\,; \qquad f'(\eta \to \infty) = 1\,. \tag{6.14}$$

[†]Since $\delta = 0$ at the leading edge of the plate, $d\delta/dx \to \infty$ as illustrated in Fig. 6.2.

The striking resemblance in the expression for δ between Eqn. 6.12 and 6.4 attests to the power of scaling analysis.

Equation 6.13 can be solved numerically using any of the standard methods. In fact, Blasius himself was successfully able to obtain the numerical solution in 1908. Since the boundary conditions Eqn. 6.14 are specified at both $\eta = 0$ and ∞, Eqn. 6.13 and its boundary conditions constitute a boundary value problem. In a numerical solution procedure, the condition at $\eta = \infty$ is usually written as follows:

$$\frac{df}{d\eta}(\eta = \eta_{max}) = 1\,,$$

where, η_{max} is some suitably large number (for boundary layer calculations η_{max}=10 is adequate). It is convenient to convert the boundary value problem into an initial value problem, owing to the availability of powerful methods such as Runge-Kutta and Runge-Kutta-Fehlberg for the solution of initial value problems. This we can do by specifying one more condition at $\eta = 0$ in place of the one at $\eta = \eta_{max}$. Since f and $df/d\eta$ are already specified at $\eta = 0$, we will choose to specify $d^2f/d\eta^2$. The value chosen for the second derivative should be such that the condition at $\eta = \eta_{max}$ is satisfied. Thus, we have a classic shooting value problem. The correct value for $f''(\eta = 0)$ can be arrived at after a few trials. The solution thus obtained is shown in Fig. 6.4.

The x-velocity u/U_∞ from Eqn. 6.10 is also plotted in Fig. 6.4. It can be seen from this figure that the edge of the boundary layer (based upon either of the criterion mentioned earlier) can be safely said to be at $\eta = 5$. Thus, the boundary thickness at any location is given as[†]

$$\delta = 5\sqrt{\frac{\nu x}{U_\infty}}\,.$$

[†]The readers may wonder if the number 5 in this expression would have been different if we had used a different value for the constant c. The edge of the boundary layer would certainly have been at a different value of η, but the product of this value with $\sqrt{2c}$ would still come out to be 5.

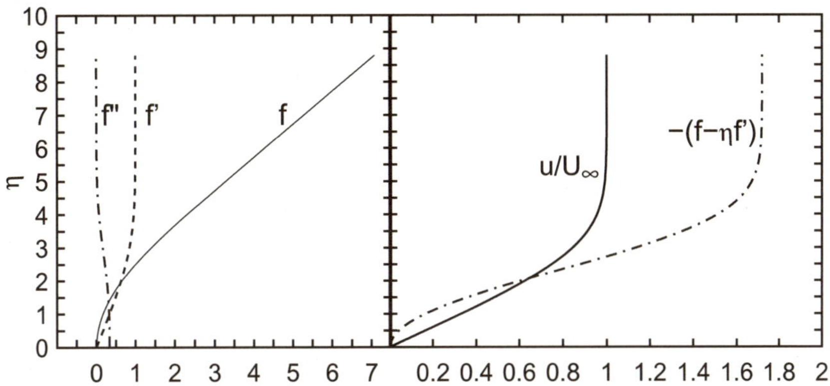

Fig. 6.4: f, f' and f'' and velocity profiles from the solution to the Blasius equation

An alternate and more useful form of this expression is,

$$\frac{\delta}{x} = \frac{5}{Re_x^{1/2}}, \tag{6.15}$$

where $Re_x = U_\infty x/\nu$ is the local Reynolds number based on the distance from the leading edge of the plate. This expression is dimensionless and more useful since all the quantities that can be varied, namely, the freestream velocity and the kinematic viscosity of the fluid, are contained nicely within the Reynolds number.

Equation 6.11 can be simplified to read as

$$v = -\frac{c\nu}{\delta(x)}\left(f - \eta f'\right) = -\frac{\nu}{2\delta(x)}\left(f - \eta f'\right).$$

This is also plotted in Fig. 6.4. As surmised earlier, v is positive throughout and it increases from zero at the wall to a maximum at the edge of the boundary layer. We note in passing that, unlike u, v is not self-similar owing to the presence of the term $\delta(x)$ in the denominator in the right hand side of the above expression.

The vorticity ω_z is given as

$$\omega_z = -\frac{\partial u}{\partial y} = -\frac{U_\infty}{\delta}\frac{\partial u}{\partial \eta} = -\frac{U_\infty}{\delta} f'' ,$$

where we have used Eqn. 6.10 in deriving the last equality. The profile of f'' plotted in Fig. 6.4 corroborates the comments made earlier regarding the variation of ω_z along the direction perpendicular to the wall.

Another quantity of interest is the drag force. The wall shear stress is given as

$$\tau_w = \mu \left.\frac{\partial u}{\partial y}\right|_{y=0} = \mu \frac{1}{\delta}\left.\frac{\partial u}{\partial \eta}\right|_{\eta=0} = \mu\sqrt{\frac{U_\infty}{\nu x}}\left.\frac{\partial u}{\partial \eta}\right|_{\eta=0} = \mu\sqrt{\frac{U_\infty}{\nu x}} U_\infty f''(0) ,$$

where we have used Eqns. 6.12 and 6.10 successively. This can be simplified to read

$$\frac{\tau_w}{\rho U_\infty^2} = \frac{f''(0)}{Re_x^{1/2}} = \frac{0.332}{Re_x^{1/2}} . \tag{6.16}$$

Note that the wall shear stress decreases as the square root of the distance from the leading edge of the plate. This is to be expected, since the fluid is decelerated from the freestream velocity to zero within a distance of δ and so the shear stress necessary to achieve this is higher when δ is smaller *i.e.* when the boundary layer is thinner and *vice versa*. It is customary to define a local skin friction coefficient, c_f, as

$$c_f = \frac{\tau_w}{\frac{1}{2}\rho U_\infty^2} = \frac{0.664}{Re_x^{1/2}} . \tag{6.17}$$

The drag force $\mathfrak{D}$ on the plate (assuming W to be the width normal to the paper) can now be evaluated as

$$\begin{aligned} \mathfrak{D} &= W \int_0^x \tau_w(\xi) d\xi & (6.18) \\ &= \rho U_\infty^2 W \int_0^x 0.332 \sqrt{\frac{\nu}{U_\infty}} \frac{1}{\xi^{1/2}} d\xi \\ &= 0.664\, \rho U_\infty^{1.5} W \nu^{0.5} x^{0.5} . \end{aligned}$$

Since the drag force is a cumulative sum starting from the leading edge, it increases monotonically as $x^{1/2}$. The above expression can be simplified to read as

$$\frac{\mathfrak{D}}{\rho U_\infty^2 W x} = \frac{0.664}{Re_x^{1/2}}. \qquad (6.19)$$

The drag force is usually expressed in dimensionless form using the coefficient of drag, C_d as

$$C_d = \frac{\mathfrak{D}}{\frac{1}{2}\rho U_\infty^2 W L} = \frac{1.328}{Re_L^{1/2}}, \qquad (6.20)$$

where L is the length of the plate. It is interesting to note from Eqns. 6.17 and 6.20 that, for a plate of length L, the coefficient of drag is twice the c_f evaluated at $x = L$ *i.e.*, $C_d = 2c_f(L)$.

6.2.2 *Integral analysis*

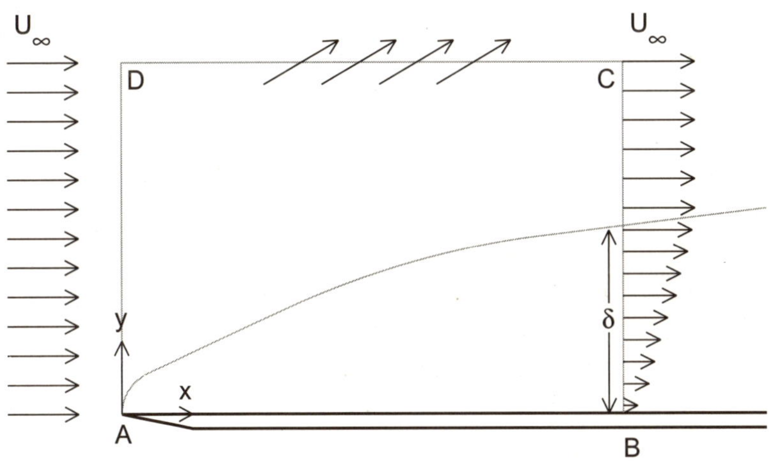

Fig. 6.5: Integral analysis of the boundary layer using the control volume ABCD.

Consider once again the boundary layer flow over a flat plate with zero pressure gradient as shown in Fig. 6.5. An integral analysis involves the application of mass and momentum conservation to a control volume such as the one marked $ABCD$ in the figure.

Since the flow is steady, the rate of change of mass in the control volume is zero. This means that the mass entering the control volume is balanced by the mass that leaves. For the control volume $ABCD$, then

$$\int_0^{y_D} \rho u \, dy - \int_0^{y_C} \rho u \, dy - \dot{m}_{CD} = 0\,,$$

where, $\dot{m}_{CD}$ is the mass flow rate through the top face of the control volume and we have assumed unit width perpendicular to the page. Note that no mass enters or leaves through the bottom face. This expression can be simplified further as

$$\dot{m}_{CD} = \int_0^{y_D} \rho U_\infty \, dy - \int_0^{\delta(x_B)} \rho u_{x=x_B} \, dy - \int_{\delta(x_B)}^{y_C} \rho U_\infty \, dy\,.$$

If we split the first integral into two parts,

$$\dot{m}_{CD} = \int_0^{\delta(x_B)} \rho U_\infty \, dy + \underline{\int_{\delta(x_B)}^{y_D} \rho U_\infty \, dy} - \int_0^{\delta(x_B)} \rho u_{x=x_B} \, dy$$
$$- \underline{\int_{\delta(x_B)}^{y_C} \rho U_\infty \, dy}\,.$$

The underlined terms in the above equation cancel each other since $y_c = y_D$ and if we combine the remaining terms, we get

$$\dot{m}_{CD} = \rho U_\infty \int_0^{\delta(x_B)} \left(1 - \frac{u_{x=x_B}}{U_\infty}\right) dy\,. \tag{6.21}$$

Note that $\dot{m}_{CD}$ is positive as explicitly illustrated in Fig. 6.5.

Conservation of momentum applied to control volume $ABCD$ requires that the rate of change of momentum in the control volume should equal the force exerted on the control volume. The net change in the x-momentum of the fluid as it passes through the control volume is given as[†]

$$-\rho \int_0^{y_D} U_\infty^2 \, dy + \rho \int_0^{y_C} u_{x=x_B}^2 \, dy + \dot{m}_{CD} U_\infty \, ,$$

where, we have used the fact that the mass that leaves the top face of the control volume does so with an x-component of velocity equal to the freestream value. Substituting for $\dot{m}_{CD}$ from Eqn. 6.21, the above expression becomes

$$-\rho \int_0^{y_D} U_\infty^2 \, dy + \rho \int_0^{y_C} u_{x=x_B}^2 \, dy + \rho \int_0^{\delta(x_B)} (U_\infty^2 - u_{x=x_B} U_\infty) \, dy \, .$$

If we split the first two integrals into two parts as before, we get

$$\underline{\underline{-\rho \int_0^{\delta(x_B)} U_\infty^2 \, dy}} \; \underline{- \rho \int_{\delta(x_B)}^{y_D} U_\infty^2 \, dy} + \rho \int_0^{\delta(x_B)} u_{x=x_B}^2 \, dy$$

$$\underline{+ \rho \int_{\delta(x_B)}^{y_C} U_\infty^2 \, dy} + \rho \int_0^{\delta(x_B)} (\underline{\underline{U_\infty^2}} - u_{x=x_B} U_\infty) \, dy \, .$$

The underlined terms in the above expression cancel each other and we are finally led to

$$-\rho \int_0^{\delta(x_B)} u_{x=x_B} (U_\infty - u_{x=x_B}) \, dy \, ,$$

as the rate of change of momentum of the fluid across the control volume. The *negative* sign indicates that there is a drag force $\mathfrak{D}$ acting on the control volume in the *negative* x-direction. Thus,

$$\mathfrak{D} = -\rho \int_0^{\delta(x_B)} u_{x=x_B} (U_\infty - u_{x=x_B}) \, dy \, . \tag{6.22}$$

[†]Note that any change in a quantity is calculated as the final value minus the initial value. For example, if the momentum of the fluid in a certain direction increases across the control volume, this would be due to the action of a force acting in that direction. Calculating the change this way ensures that the directions are consistent.

However, from Eqn. 6.18, it is easy to see that $\tau_w(x) = -d\mathfrak{D}/dx$ [†] and so

$$\tau_w(x_B) = \rho U_\infty^2 \frac{d}{dx} \int_0^{\delta(x_B)} \frac{u_{x=x_B}}{U_\infty} \left(1 - \frac{u_{x=x_B}}{U_\infty}\right) dy\,.$$

If we substitute for τ_w, we are finally led to

$$\mu \left.\frac{\partial u}{\partial y}\right|_{\substack{x = x_B \\ y=0}} = \rho U_\infty^2 \frac{d}{dx} \int_0^{\delta(x_B)} \frac{u_{x=x_B}}{U_\infty} \left(1 - \frac{u_{x=x_B}}{U_\infty}\right) dy\,. \quad (6.23)$$

We note that if a reasonable guess for the velocity profile at $x = x_B$ were available, this would allow both sides of Eqn. 6.23 to be evaluated, leading to an equation for $\delta(x_B)$. Three possible choices for the velocity profile are:

$$\frac{u}{U_\infty} = \begin{cases} \dfrac{y}{\delta} \\[2ex] \dfrac{2y}{\delta} - \dfrac{y^2}{\delta^2} \\[2ex] \dfrac{3y}{\delta} - \dfrac{3y^2}{\delta^2} + \dfrac{y^3}{\delta^3} \end{cases} \quad (6.24)$$

All the profiles satisfy $u = 0$ at $y = 0$ and $u = U_\infty$ at $y = \delta$. The second profile satisfies, in addition, $\partial u/\partial y = 0$ at $y = \delta$. The third profile, satisfies, $\partial^2 u/\partial y^2 = 0$ at $y = \delta$ also. Obviously, higher order polynomials in y/δ are also possible. Furthermore, the profiles need not only be polynomials in y/δ. Trigonometric and transcendental functions in y/δ can also be used. It must also be noted that the guessed profiles are self-similar since they are all functions of y/δ.

The development leading up to Eqn. 6.23 has been exact in that no approximations have been made. The use of a guessed velocity profile

[†]The negative sign has been added since we are considering the drag on the control volume - not the plate.

makes the analysis approximate from here onwards. Substitution of the quadratic velocity profile from Eqn. 6.24 into Eqn. 6.23 results in

$$\frac{2\mu U_\infty}{\delta} = \rho U_\infty^2 \frac{d}{dx}\left(\frac{2\delta}{15}\right) = \rho U_\infty^2 \frac{2}{15}\frac{d\delta}{dx},$$

where the subscript x_B has been dropped as there is no danger of ambiguity. Therefore,

$$\delta \frac{d\delta}{dx} = \frac{15\nu}{U_\infty}.$$

This can be integrated to yield

$$\delta^2 = \frac{30\nu x}{U_\infty} \quad \text{or} \quad \frac{\delta}{x} = \frac{5.5}{Re_x^{1/2}}.$$

This compares quite well with the expression given in Eqn. 6.15 obtained using the exact analysis. In addition,

$$\frac{\tau_w}{\rho U_\infty^2} = \frac{0.3636}{Re_x^{1/2}}$$

which also compares reasonably well with the result (Eqn. 6.16) from the exact analysis. Use of higher order polynomials will yield even better results.

6.2.3 *Displacement and momentum thickness*

The integrals in the right hand side of Eqns. 6.21 and 6.23 have a special significance in the context of boundary layers. This is best understood by studying the streamlines corresponding to the flow shown in Fig. 6.2. These are shown in Fig. 6.6(a). The streamlines are uniformly spaced as they approach the plate and begin to diverge once inside the boundary layer. The streamlines move apart owing to the deceleration of the flow inside the boundary layer. The spacing between the streamlines in Fig. 6.6(a) clearly shows that the deceleration is high near the wall and diminishes

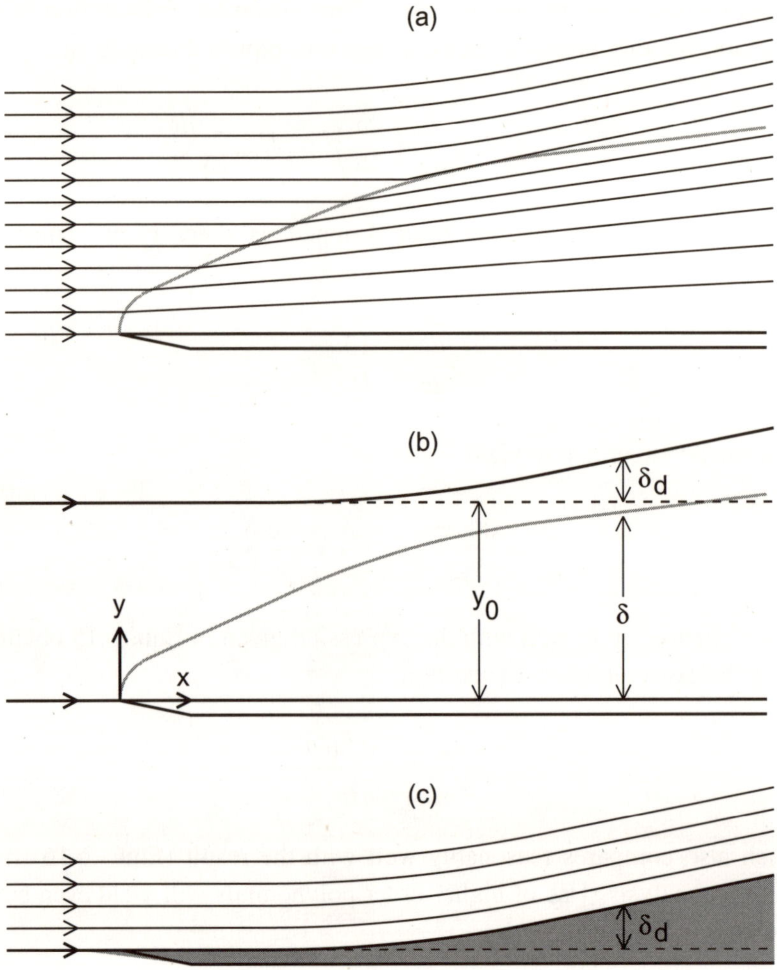

Fig. 6.6: (a) Streamlines for boundary layer flow over a flat plate (b) Illustration of the displacement thickness and (c) Potential flow over a flat plate with boundary layer effect taken into account

towards the edge of the boundary layer. The streamlines are parallel to each other outside the boundary layer, although they are displaced in the y-direction. The small, but non-zero, slope of the streamlines indicates that the y-component of velocity inside the boundary layer is small but non-zero. The streamlines in the freestream shown in Fig. 6.6(a) indicate that this is true in the frestream also.

Consider a streamline that is initially at a distance y_0 from the streamline at $y = 0$ as shown in Fig. 6.6(b). This streamline is assumed to lie entirely in the freestream *i.e.,* it never enters the boundary layer. However, it "feels" the presence of the boundary layer, since it is displaced upwards by δ_d. This displacement can be calculated by using the fact that the mass flow rate between this streamline and the one along $y = 0$ has to be the same everywhere. Accordingly, the mass flow rate between these two streamlines at an $x = constant$ location ahead of the plate is simply $\rho U_\infty y_0$ (assuming unit width normal to the page). If we equate this with the mass flow rate at any other section downstream of the leading edge of the plate, we get

$$\begin{aligned}
\rho U_\infty y_0 &= \rho \int_0^{y_0+\delta_d} u dy \\
&= \rho \left[\int_0^{\delta} u dy + \int_{\delta}^{y_0} u dy + \int_{y_0}^{y_0+\delta_d} u dy \right] \\
&= \rho \left[\int_0^{\delta} u dy + U_\infty (y_0 - \delta) + U_\infty \delta_d \right] \\
&= \rho \int_0^{\delta} (u - U_\infty) dy + \rho U_\infty y_0 + \rho U_\infty \delta_d \,.
\end{aligned}$$

Upon simplification, this leads to

$$\delta_d = \int_0^{\delta} \left(1 - \frac{u}{U_\infty} \right) dy \,. \tag{6.25}$$

The quantity δ_d is called the displacement thickness of the boundary layer. We can substitute $u/U_\infty = f'$ from Eqn. 6.10 and then evaluate the integral numerically. This gives $\delta_d = 0.3442\delta$. Alternatively, if we use the quadratic velocity profile from Eqn. 6.24, we get $\delta_d = \delta/3$.

The physical significance of the displacement thickness is illustrated in Fig. 6.6(c). Since the net effect of the boundary layer on the *undisturbed freestream* is to displace it by δ_d, the flow field shown in Fig. 6.6(c) results. Of course, the streamlines for potential flow over a flat plate with zero pressure gradient will simply be horizontal lines, if the effect of the boundary layer is not accounted for. However, accounting for the boundary layer effect changes the streamlines in the manner illustrated in Fig. 6.6(c). We thus see an interaction between the boundary layer and the potential flow. The interaction is weak, however, since it results in only a slight change in the potential flow.

Displacement thickness is a very useful concept in practical applications. An example of such an application is the wind or water tunnel†. The tunnel allows experimental investigations of the flow field around objects when placed in a freestream, to be carried out, by providing a potential flow in the test section where the object is placed. In order for the experimental results to be useful, the extent of the potential flow region in the test section must be quite large when compared to the size of the object being tested. In other words, the object must not "feel" the presence of the confining walls of the tunnel. However, the growing boundary layers on the walls of the tunnels make this requirement difficult to achieve. The displacement thickness comes in very useful in this situation, since we can replace the walls of the tunnel in a manner illustrated in Fig. 6.6(c), and determine quantitatively the blockage created by the boundary layers. A well designed tunnel, usually, has a blockage ratio (ratio of the cross-sectional area of the object being tested to the cross-sectional area of the test section) of 1% or less.

It has been already mentioned that the fluid loses momentum continuously owing to the drag exerted by the plate. The effect of this momentum deficit on the freestream can be quantified through a momentum thickness. Consider once again the streamline that is initially at a distance y_0 from the streamline at $y = 0$ as shown in Fig. 6.6(b). The momentum that crosses an $x = constant$ location ahead of the plate, between these two streamlines is simply $\rho U_\infty^2 y_0$ (assuming unit width normal to the page). The momentum that crosses at some other $x = constant$ location downstream of the

† Since the kinematic viscosity of water is less than that of air, the thickness of the boundary layer, for a given freestream velocity, is less in water than in air, as Eqn. 6.15 shows.

leading edge of the plate is

$$
\begin{aligned}
&= \rho \int_0^{y_0+\delta_d} u^2 dy \\
&= \rho \left[\int_0^{\delta} u^2 dy + \int_{\delta}^{y_0} u^2 dy + \int_{y_0}^{y_0+\delta_d} u^2 dy \right] \\
&= \rho \left[\int_0^{\delta} u^2 dy + U_\infty^2 (y_0 - \delta) + U_\infty^2 \delta_d \right] \\
&= \rho \int_0^{\delta} (u^2 - U_\infty^2) dy + \rho U_\infty^2 y_0 + \rho U_\infty^2 \delta_d \\
&= \rho U_\infty^2 y_0 + \rho \int_0^{\delta} (u^2 - \underline{U_\infty^2}) dy + \rho U_\infty \int_0^{\delta} (\underline{U_\infty} - u) dy \\
&= \rho U_\infty^2 y_0 - \rho \int_0^{\delta} u(U_\infty - u) dy \,,
\end{aligned}
$$

where, we have used Eqn. 6.25. Also, the underlined terms cancel each other. The second term represents the momentum deficit at a given $x = constant$ location. If we write

$$
\rho U_\infty^2 \delta_m = \rho \int_0^{\delta} u(U_\infty - u) dy \,,
$$

we can interpret δ_m as the thickness of the layer in the freestream that would carry the same momentum. Since this represents a momentum deficit, the drag exerted by the plate, in effect, results in the removal of a layer of fluid of thickness δ_m from the freestream (or, alternatively, an added plate thickness of δ_m) at each location downstream of the leading edge. The momentum boundary layer thickness δ_m is thus defined as

$$
\delta_m = \int_0^{\delta} \frac{u}{U_\infty} \left(1 - \frac{u}{U_\infty} \right) dy \,. \tag{6.26}
$$

If we evaluate the momentum thickness using the Blasius solution, we get $\delta_m = 0.1328\delta$. On the other hand, if we use the quadratic profile from Eqn. 6.24, we get $\delta_m = (2/15)\delta$. Note that $\delta > \delta_d > \delta_m$.

In addition to the displacement and momentum thicknesses, an energy thickness is also customarily defined. The interested reader is urged to consult the books suggested at the end for information on this topic.

6.3 Boundary layer flows with non-zero pressure gradient

When the freestream pressure gradient is non-zero, then Bernoulli's equation shows that $dU_\infty/dx \neq 0$. This is the only modification that needs to be made to the analysis of the previous section.

6.3.1 *Falkner-Skan similarity solutions*

We start from the same point as in Section 6.2.1. With the definition for η remaining the same, it is easy to see that there is no change in Eqn. 6.10. Equation 6.11 has to be modified. Accordingly,

$$\begin{aligned} v = -\frac{\partial \psi}{\partial x} &= -\left(U_\infty f \,\frac{d\delta}{dx} + U_\infty \delta \,\frac{\partial f}{\partial x} + f\delta \,\frac{dU_\infty}{dx} \right) \\ &= -\left(U_\infty f \,\frac{d\delta}{dx} + U_\infty \delta \,\frac{\partial \eta}{\partial x}\frac{df}{d\eta} + f\delta \,\frac{dU_\infty}{dx} \right) \\ &= -f\delta \,\frac{dU_\infty}{dx} - U_\infty \,\frac{d\delta}{dx}\left(f - \eta f'\right). \end{aligned} \tag{6.27}$$

Substitution of Eqns. 6.10 and 6.27 into Eqn. 6.7 gives

$$U_\infty f' \frac{\partial}{\partial x}(U_\infty f') - \left[f\delta \,\frac{dU_\infty}{dx} + U_\infty \,\frac{d\delta}{dx}(f - \eta f') \right] \frac{\partial}{\partial y}(U_\infty f')$$
$$= U_\infty \frac{dU_\infty}{dx} + \nu \frac{\partial^2}{\partial y^2}(U_\infty f').$$

If we carry the differentiation through, we get

$$U_\infty f' \left(\frac{dU_\infty}{dx} f' - U_\infty \,\frac{\eta}{\delta}\,\frac{d\delta}{dx} f'' \right)$$

$$- \left[f\delta \frac{dU_\infty}{dx} + U_\infty \frac{d\delta}{dx}(f - \underline{\eta f'}) \right] \left(U_\infty \frac{1}{\delta} f'' \right)$$
$$= U_\infty \frac{dU_\infty}{dx} + \nu \frac{U_\infty}{\delta^2} f''' \,.$$

After cancelling the underlined terms, this equation simplifies to

$$\frac{dU_\infty}{dx} f'^2 - \left(\frac{dU_\infty}{dx} + \frac{U_\infty}{\delta} \frac{d\delta}{dx} \right) f f'' = \frac{dU_\infty}{dx} + \frac{\nu}{\delta^2} f''' \,.$$

Upon rearrangement, we get

$$f''' + \frac{\delta}{\nu} \frac{d}{dx}(U_\infty \delta) f f'' + \left(\frac{\delta^2}{\nu} \frac{dU_\infty}{dx} \right) (1 - f'^2) = 0 \,.$$

Or

$$f''' + \alpha f f'' + \beta(1 - f'^2) = 0 \,, \tag{6.28}$$

where

$$\frac{\delta}{\nu} \frac{d}{dx}(U_\infty \delta) = \alpha \quad \text{and} \quad \frac{\delta^2}{\nu} \frac{dU_\infty}{dx} = \beta \,. \tag{6.29}$$

Boundary conditions remain the same as in Eqn. 6.14. Since we are looking for a self-similar solution to Eqn. 6.28, the coefficients α and β have to be constants. Therefore, it becomes clear that only those potential flows that have a certain form for U_∞ can have a self-similar boundary layer solution. Two such solutions, for which the potential flow solutions were presented in the previous chapter, are discussed next. We note in passing that, the combination $\alpha = 1/2$ and $\beta = 0$ identically recovers the Blasius solution.

The expression for the wall shear stress in this case is very similar to Eqn. 6.16,

$$\frac{\tau_w}{\rho U_\infty^2} = \frac{f''(0)}{Re_x^{1/2}} \,. \tag{6.30}$$

6.3.1.1 *Flow over a wedge*

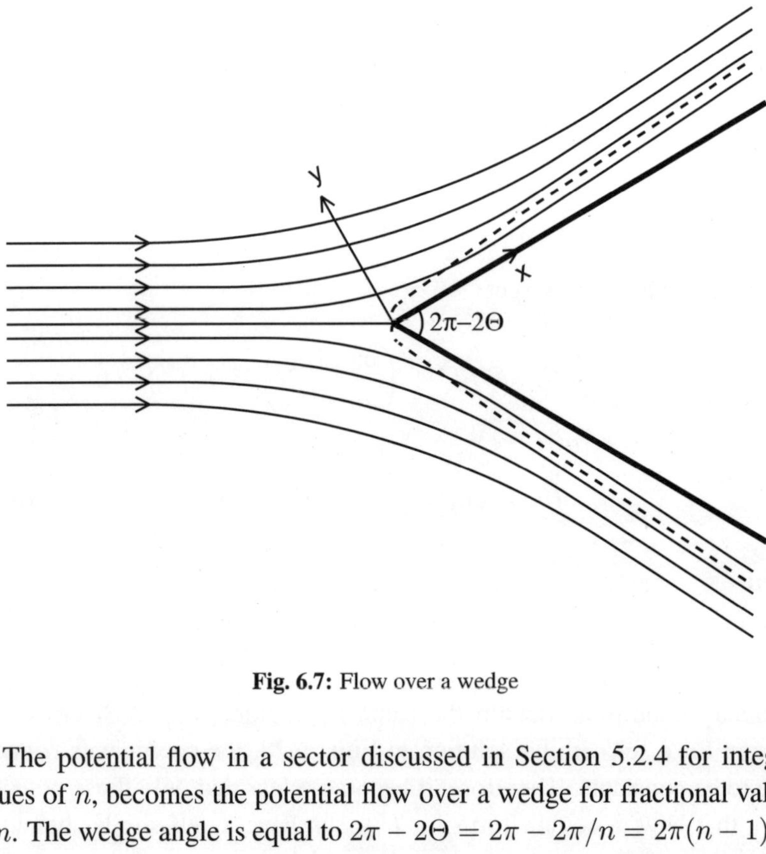

Fig. 6.7: Flow over a wedge

The potential flow in a sector discussed in Section 5.2.4 for integral values of n, becomes the potential flow over a wedge for fractional values of n. The wedge angle is equal to $2\pi - 2\Theta = 2\pi - 2\pi/n = 2\pi(n-1)/n$. For example, $n = 2$ corresponds to the potential flow in a 60° sector (Fig. 5.22), whereas $n = 6/5$, corresponds to the potential flow over a 60° wedge. The outer potential flow as well as the boundary layer flow over the wedge are illustrated in Fig. 6.7. The boundary layer on each face of the wedge is shown as a dashed line in this figure. We know from Eqn. 5.34, that the radial velocity on the surface of the wedge is equal to $u_r = nr^{n-1}$ and the tangential component is zero. With the x, y coordinate system oriented as shown in Fig. 6.7, the freestream velocity for the boundary layer calculation is thus[†]

[†]It must be recalled that the matching condition is applied with the boundary layer coordinate tending to ∞ and the outer coordinate tending to 0.

$$U_\infty = nx^{n-1}\,.$$

If a self-similar boundary layer solution is admissible for this freestream velocity distribution, then α and β must be constants. We proceed to evaluate these. Since $dU_\infty/dx = n(n-1)x^{n-2}$, we can get

$$\delta^2 = \frac{\beta\nu}{n(n-1)}\frac{1}{x^{n-2}}$$

from the second relation in Eqn. 6.29. Therefore

$$\delta U_\infty = \sqrt{\frac{\beta\nu}{n(n-1)}}\,\frac{1}{x^{(n-2)/2}}\,nx^{n-1} = \sqrt{\frac{n\beta\nu}{n-1}}\,x^{n/2}\,,$$

and

$$\frac{d}{dx}(\delta U_\infty) = \sqrt{\frac{n\beta\nu}{n-1}}\,\frac{n}{2}\,x^{(n-2)/2}\,.$$

It follows then that

$$\delta\frac{d}{dx}(\delta U_\infty) = \sqrt{\frac{\beta\nu}{n(n-1)}}\,\frac{1}{x^{(n-2)/2}}\sqrt{\frac{n\beta\nu}{n-1}}\,\frac{n}{2}\,x^{(n-2)/2} = \frac{\beta\nu}{n-1}\,\frac{n}{2}\,.$$

This, of course, is equal to $\alpha\nu$ from the first relation in Eqn. 6.29. Therefore

$$\alpha = \frac{\beta}{n-1}\,\frac{n}{2}\,.$$

If we let $\beta = 2(n-1)/n$, then $\alpha = 1$. A self-similar solution is now possible. With this choice for β, the wedge angle becomes equal to $\pi\beta^\dagger$. Also

$$\delta = \frac{\sqrt{2\nu}}{n}\,\frac{1}{x^{n/2-1}}\,.$$

†The simplicity and beauty of the expression for the wedge angle demonstrates the rationale behind the choice for β.

With the above choice for α and β, Eqn. 6.28 becomes

$$f''' + ff'' + \beta(1 - f'^2) = 0\,.$$

This can be solved numerically in the same manner as the Blasius equation for a given value of n (or, alternatively β) to obtain the complete boundary layer solution. For example, $n = 6/5$ ($\beta = 1/3$) corresponds to flow over a 60° wedge.

6.3.1.2 *Plane stagnation (Hiemenz) flow*

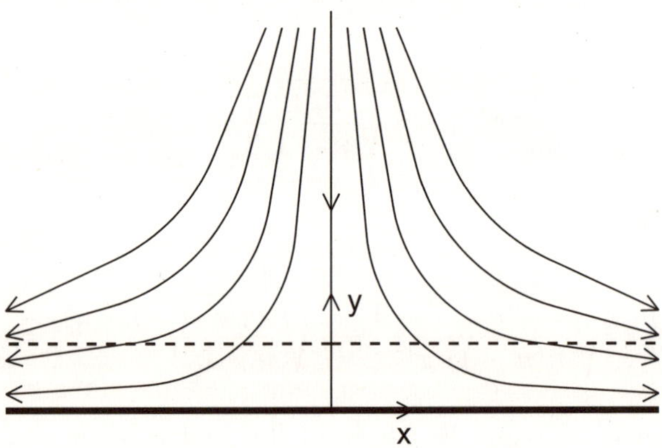

Fig. 6.8: 2D plane stagnation (Hiemenz) flow

The potential flow towards a wall (180° sector) was discussed in Section 5.2.4 and illustrated in Fig. 5.23. It was mentioned that the potential flow slips over the surface. In reality, a boundary layer develops adjacent to the wall on either side of the stagnation streamline. In this case also, the flow inside the boundary layer exhibits self-similarity. The complete solution for this problem was first obtained by Hiemenz.

The plane stagnation flow can be treated as a special case (n=2) of the flow over a wedge discussed above. With this value for n, we have $U_\infty = 2x$,

$\beta = 1$ and $\delta = \sqrt{\nu/2}$. Equation 6.28 becomes

$$f''' + ff'' + (1 - f'^2) = 0\,.$$

Numerical solution to this equation is given in *Boundary Layer Theory* by Schlichting & Gersten. The remarkable feature of this flow field is that the boundary thickness is a constant! The flow field is illustrated in Fig. 6.8. The boundary layer is shown in this figure using a dashed line. In contrast to Fig. 5.23, the increase in the spacing between the streamlines owing to the deceleration inside the boundary layer can be seen in Fig. 6.8.

6.3.1.3 *Flow over a flat plate with a sink at the origin*

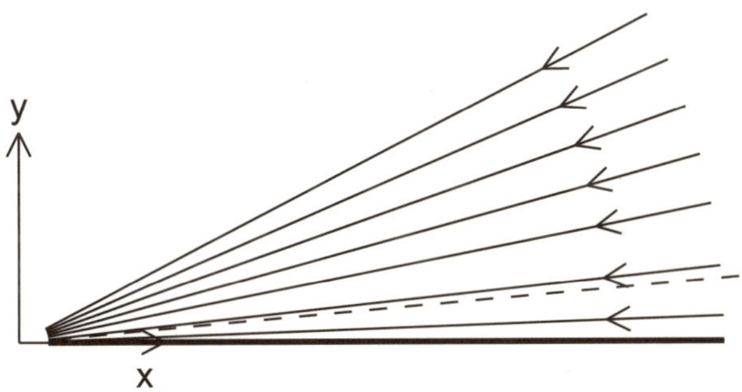

Fig. 6.9: Boundary layer flow over a flat plate with a sink at the origin

Potential flow due to a sink placed at the origin was discussed in Section 5.2.1. The velocity of the fluid for this case is $u_r = -q/(2\pi r)$ with the azimuthal velocity being zero everywhere (Eqn. 5.22). Now consider the flow illustrated in Fig. 6.9 where a flat plate (or wall) is placed along the x-axis such that the sink is just ahead of the leading edge of the plate. A boundary layer develops along the wall as shown in the figure. The deceleration of the flow in the boundary layer is also evident from this figure. With x, y axes as indicated in the figure, the freestream velocity for the boundary layer flow is $U_\infty = -q/(2\pi x)$. From Eqn. 6.29,

$$\delta^2 \frac{q}{2\pi x^2} = \beta \;\Rightarrow\; \delta = \sqrt{\frac{2\pi\beta}{q}}\, x\,.$$

The boundary layer thickness varies linearly with x. Therefore $d(U_\infty \delta)/dx = 0$ and hence $\alpha = 0$. The value for β is arbitrary. Equation 6.28 thus becomes

$$f''' + \beta(1 - f'^2) = 0\,.$$

We note in passing that, a similar solution with a source instead of a sink is not possible.

6.3.2 *von Kármán - Pohlhausen integral solution*

We start by integrating the continuity equation 6.1 along the y-direction,

$$\int_0^y \frac{\partial u}{\partial x}\, d\xi \,+\, \int_0^y \frac{\partial v}{\partial \xi}\, d\xi = 0\,,$$

where ξ is a dummy integration variable. The second integration can be carried through to give

$$\int_0^y \frac{\partial u}{\partial x}\, d\xi \,+\, v|_{\xi=0}^{y} = 0\,.$$

Since $v(y = 0) = 0$, we get

$$v = -\int_0^y \frac{\partial u}{\partial x}\, d\xi\,. \tag{6.31}$$

If we now integrate the momentum equation 6.7, between $y = 0$ and δ, we get

$$\int_0^\delta u \frac{\partial u}{\partial x}\, dy + \int_0^\delta v \frac{\partial u}{\partial y}\, dy = \int_0^\delta U_\infty \frac{dU_\infty}{dx}\, dy + \nu \int_0^\delta \frac{\partial^2 u}{\partial y^2}\, dy\,.$$

Upon rearranging, and integrating the viscous term, we get

$$\int_0^\delta u \frac{\partial u}{\partial x} dy + \int_0^\delta v \frac{\partial u}{\partial y} dy - \int_0^\delta U_\infty \frac{dU_\infty}{dx} dy = \nu \left. \frac{\partial u}{\partial y} \right|_0^\delta . \tag{6.32}$$

Since $\partial u/\partial y = 0$ at $y = \delta$, the right hand side of this equation is simply equal to $-\tau_w/\rho$. Let us now simplify the second term in the above equation by first substituting for v from Eqn. 6.31,

$$\int_0^\delta v \frac{\partial u}{\partial y} dy = -\int_0^\delta \underline{\left(\int_0^y \frac{\partial u}{\partial x} d\xi\right)} \; \underline{\frac{\partial u}{\partial y} dy} .$$

If we integrate by parts with the terms grouped as shown, we get

$$\int_0^\delta v \frac{\partial u}{\partial y} dy = -\left[\left(u \int_0^y \frac{\partial u}{\partial x} d\xi\right)_0^\delta - \int_0^\delta u \frac{\partial u}{\partial x} dy\right] .$$

This can be simplified to read

$$\int_0^\delta v \frac{\partial u}{\partial y} dy = -U_\infty \int_0^\delta \frac{\partial u}{\partial x} dy + \int_0^\delta u \frac{\partial u}{\partial x} dy .$$

Upon substituting this in Eqn. 6.32 and collecting terms, we get

$$\int_0^\delta \left(2u \frac{\partial u}{\partial x} - U_\infty \frac{\partial u}{\partial x} - U_\infty \frac{dU_\infty}{dx}\right) dy = -\frac{\tau_w}{\rho} .$$

This can be written as

$$\frac{\partial}{\partial x} \int_0^\delta [u(U_\infty - u)] \, dy + \frac{dU_\infty}{dx} \int_0^\delta (U_\infty - u) \, dy = \frac{\tau_w}{\rho} ,$$

where, we have taken the partial derivative outside the integral, since the integration is with respect to y. If we now use the definition of the displacement thickness and the momentum thickness, we get

$$\frac{d}{dx}\left(U_\infty^2 \delta_m\right) + U_\infty \frac{dU_\infty}{dx}\,\delta_d = \frac{\tau_w}{\rho}\,.$$

The partial derivative has been replaced with an ordinary derivative, since the integral is a definite integral in y, and after integration, x is the only variable left. The above equation can also be written as

$$U_\infty^2 \frac{d\delta_m}{dx} + (2\delta_m + \delta_d) U_\infty \frac{dU_\infty}{dx} = \frac{\tau_w}{\rho}\,. \tag{6.33}$$

This is the momentum integral equation for a 2D, incompressible boundary layer, first derived by von Kármán. This equation becomes identical to Eqn. 6.23, when $dU_\infty/dx = 0$, as it should.

No approximations have been used in the development of Eqn. 6.33. The analysis proceeds from here onwards with an assumed velocity profile that was proposed by Pohlhausen. This profile is given as

$$\frac{u}{U_\infty} = a\eta + b\eta^2 + c\eta^3 + d\eta^4\,, \quad 0 \le \eta \le 1\,,$$

where $\eta = y/\delta(x)$, as usual. This profile automatically satisfies the condition that $u = 0$ at $y = 0$. Pohlhausen used the following boundary conditions to evaluate the constants a, b, c and d in the profile.

$$\nu \left.\frac{\partial^2 u}{\partial y^2}\right|_{y=0} = -U_\infty \frac{dU_\infty}{dx}\,,$$

$$u\big|_{y=\delta} = U_\infty\,,$$

$$\left.\frac{\partial u}{\partial y}\right|_{y=\delta} = 0\,,$$

$$\left.\frac{\partial^2 u}{\partial y^2}\right|_{y=\delta} = 0\,.$$

The first of the above conditions has been obtained by applying Eqn. 6.7 at $y = 0$. The above conditions involving y can be converted into conditions involving η as follows:

$$\left.\frac{\partial^2(u/U_\infty)}{\partial\eta^2}\right|_{\eta=0} = -\frac{\delta^2}{\nu}\frac{dU_\infty}{dx}\,,$$

$$\left.\frac{u}{U_\infty}\right|_{\eta=1} = 1\,,$$

$$\left.\frac{\partial(u/U_\infty)}{\partial y}\right|_{\eta=1} = 0\,,$$

$$\left.\frac{\partial^2(u/U_\infty)}{\partial y^2}\right|_{\eta=1} = 0\,.$$

If we let

$$\Lambda(x) = \frac{\delta^2}{\nu}\frac{dU_\infty}{dx}\,, \tag{6.34}$$

then $a = 2 + \Lambda/6$, $b = -\Lambda/2$, $c = -2 + \Lambda/2$ and $d = 1 - \Lambda/6$. The quantity Λ is usually called as shape factor. It has the same sign as dU_∞/dx, since δ^2 is always positive. Hence, Λ is positive for flows in which U_∞ increases with x, *i.e.,* an accelerating freestream. It follows from the Bernoulli equation that, the freestream pressure gradient dp_∞/dx is negative, *i.e.,* favorable, in such flows. Conversely, Λ is negative for flows with a decelerating freestream or adverse pressure pressure gradient. Also, note that Λ is a function of x, in general.

Pohlhausen velocity profiles for different values of Λ are shown in Fig. 6.10. The case $\Lambda = 0$, of course, corresponds to flow over a flat plate with zero pressure gradient. When the pressure gradient is favorable and strong enough that $\Lambda > 12$, the dimensionless velocity u/U_∞ inside the boundary layer itself exceeds the freestream value, which is physically not possible. On the other hand, when the pressure gradient is strong and adverse, with $\Lambda < -12$, then u/U_∞ becomes negative near the wall. Although this

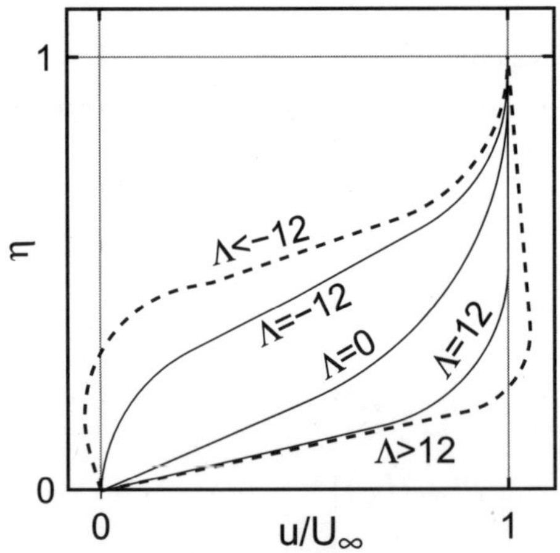

Fig. 6.10: Boundary layer velocity profiles for different values of the parameter Λ

is possible when the boundary layer separates, and an adverse pressure gradient does lead to separation, the theory developed so far does not have the capability to reliably predict separation. Hence, the theory is useful when $-12 \leq \Lambda \leq 12$.

It is straightforward to show that

$$\frac{\delta_d}{\delta} = \frac{3}{10} - \frac{\Lambda}{120}$$

$$\frac{\delta_m}{\delta} = \frac{37}{315} - \frac{\Lambda}{945} - \frac{\Lambda^2}{9072} \tag{6.35}$$

$$\tau_w = \mu \left.\frac{\partial u}{\partial y}\right|_{y=0} = \frac{\mu U_\infty}{\delta} \left.\frac{\partial (u/U_\infty)}{\partial \eta}\right|_{\eta=0} = \frac{\mu U_\infty}{\delta}\left(2 + \frac{\Lambda}{6}\right).$$

Substitution of the above expressions into Eqn. 6.33 will lead to a differential equation involving Λ, or, equivalently, δ (it must be kept in mind that U_∞ and dU_∞/dx are both known). It can be shown[†] that this differential equation eventually leads to the following result:

[†]These details are skipped here for the sake of brevity. Interested readers can consult the books in the suggested reading for the details.

$$\delta_m^2 = \frac{0.47\nu}{U_\infty^6} \int_0^x U_\infty^5(\xi) d\xi \, , \tag{6.36}$$

where ξ is a dummy integration variable. Note that the right hand side is completely known. With δ_m known, Λ can be determined from Eqns. 6.34 and 6.35 as

$$\left(\frac{37}{315} - \frac{\Lambda}{945} - \frac{\Lambda^2}{9072} \right)^2 \Lambda = \frac{\delta_m^2}{\nu} \frac{dU_\infty}{dx} \, . \tag{6.37}$$

To summarize, once the potential flow solution *i.e.*, $U_\infty(x)$ is known, δ_m can be determined from Eqn. 6.36. Λ can then be obtained by solving Eqn. 6.37. Once Λ is known, all other quantities of interest can be calculated. This is demonstrated below for some of the potential flow solutions that were discussed earlier.

6.3.2.1 *Flow over a flat plate*

The solution for the boundary layer over a flat plate with zero pressure gradient ($\Lambda = 0$) is discussed first, since this is the easiest. Although the solution using an integral analysis has already been presented, it is worthwhile revisiting as the Pohlhausen velocity profile is better than the polynomial profiles used earlier (Eqn. 6.24) and hence can be expected to give better results.

Since U_∞ is a constant, Eqn. 6.36 gives

$$\delta_m = \sqrt{\frac{0.47\nu x}{U_\infty}} \quad \text{or} \quad \frac{\delta_m}{x} = \frac{0.686}{Re_x^{1/2}} \, .$$

From Eqn. 6.35, with $\Lambda = 0$, we get

$$\frac{\delta}{x} = \frac{5.84}{Re_x^{1/2}} \, .$$

This expression, contrary to our expectation, compares slightly worse than before with Eqn. 6.15. Also from Eqn. 6.35,

$$\frac{\tau_w}{\rho U_\infty^2} = \frac{0.3426}{Re_x^{1/2}} .$$

This expression compares better with Eqn. 6.16.

6.3.2.2 *Flow over a wedge*

As already mentioned, $U_\infty = nx^{n-1}$ for this case. If we substitute this into Eqn. 6.36 and carry the integration through, we get

$$\delta_m^2 = \frac{0.47\nu}{U_\infty^6} \frac{n^5 x^{5n-4}}{5n-4} .$$

Upon simplification, this becomes

$$\delta_m^2 = \frac{0.47\nu}{5n-4} \frac{1}{x^{n-2}} .$$

If we substitute this into Eqn. 6.37, we are led to

$$\left(\frac{37}{315} - \frac{\Lambda}{945} - \frac{\Lambda^2}{9072} \right)^2 \Lambda = \frac{0.47\nu}{5n-4} \frac{1}{x^{n-2}} \frac{1}{\nu} n(n-1)x^{n-2}$$
$$= \frac{0.47n(n-1)}{5n-4} .$$

Let us now select $n = 6/5$, which corresponds to the flow over a 60° wedge. Thus

$$\delta_m = \sqrt{0.235\nu}\, x^{2/5} = 0.485\sqrt{\nu}\, x^{2/5}$$

and

$$\left(\frac{37}{315} - \frac{\Lambda}{945} - \frac{\Lambda^2}{9072} \right)^2 \Lambda = 0.0564 .$$

Since $dU_\infty/dx > 0$ for this flow, the pressure gradient is favorable and thus $\Lambda > 0$. Solving the above equation, we get $\Lambda = 4.6475$. All other quantities of interest can be determined now. For example,

$$\delta = 4.403\sqrt{\nu}\, x^{2/5}$$

and

$$\frac{\tau_w}{\rho U_\infty^2} = 0.525\sqrt{\nu}\, x^{1/5}\,.$$

Note that the functional form for δ obtained here agrees exactly with that obtained using the Falkner-Skan similarity principle.

6.4 Separation and drag

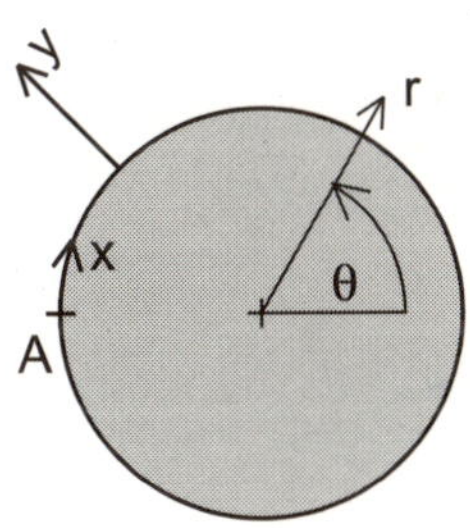

Fig. 6.11: Coordinate system (x, y) for the boundary layer calculation for flow over a circular cylinder

Consider the flow over a circular cylinder with zero circulation (Fig. 5.14) again. A boundary layer calculation to fix the potential flow solution would use the x, y coordinate system shown in Fig. 6.11, where x runs along the surface of the cylinder† and y is normal to the surface of the cylinder. The origin for this coordinate system would be at the front stagnation point A. In the previous chapter, the velocity on the surface of the cylinder was shown to be $u_r = 0$ and $u_\theta = -2U_\infty \sin\theta$, where U_∞ is

†Although the x coordinate runs along the surface of the cylinder, Eqns. 6.1 and 6.7 are still applicable so long as the boundary layer thickness $\delta \ll a$. In this limit, curvature effects inside the boundary layer can be ignored.

the freestream velocity far away from the cylinder. The freestream velocity for the boundary layer calculation is thus $U = 2U_\infty \sin(x/a)$, where a is the radius of the cylinder. The negative sign disappears since x and θ run in opposite directions. It is easy to see that U increases from $x = 0$ till $x = \pi a/2$ (or $\theta = \pi/2$) and decreases thereafter. Equation 5.30 shows that the pressure gradient is favorable till $x = \pi a/2$ and becomes adverse afterwards. The deceleration of the potential flow outside the boundary layer and the effect of the adverse pressure gradient on the boundary layer are illustrated in Fig. 6.12.

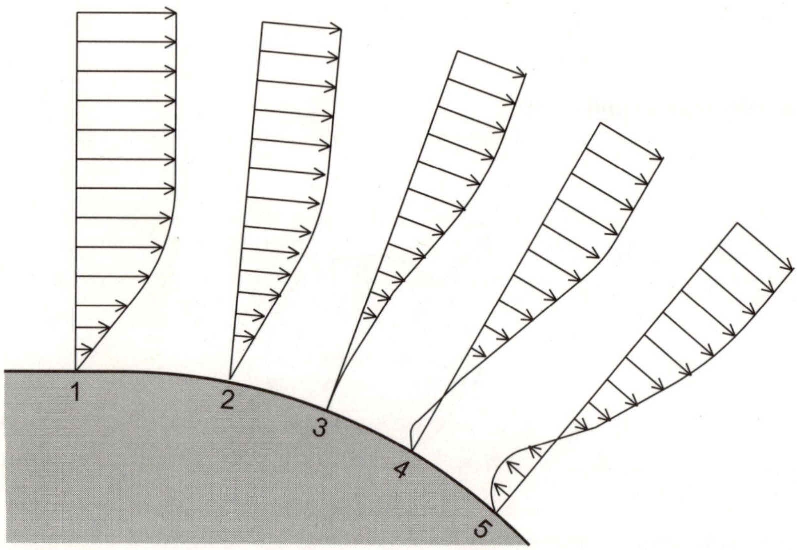

Fig. 6.12: Separation of the boundary layer under an adverse pressure gradient

As already mentioned, the surface continuously extracts momentum from the flow causing the flow in the boundary layer to decelerate continuously. This, in turn, causes the boundary layer to grow. This can be seen from the velocity profiles at stations 1 and 2 in Fig. 6.12. Eventually this causes the fluid near the surface to stagnate, as shown at station 3 in Fig. 6.12. Note that, at station 3, $\partial u/\partial y = 0$ at the wall. Once this happens, the boundary layer separates from the surface and is lifted up. A separated flow region containing slow moving fluid with a negative velocity near the surface takes its place. This change in the flow field is usually large enough to alter the potential flow in the outer region. The potential flow will no longer be as shown in Fig. 5.14 and the flow loses its symmetry about the vertical centerline of the cylinder. Thus, the boundary layer equations are no longer valid beyond the point of separation.

Left to itself, separation of the boundary layer is inevitable, even when the pressure gradient is favorable. This is because the the fluid close to the surface at some distance from the leading edge (or $x = 0$), has been decelerated so much that it does not possess enough momentum to proceed further. An adverse pressure gradient causes separation to occur soon, whereas, a favorable pressure gradient delays it. In fact, for laminar flow over a cylinder, experiments (and numerical simulations of the full Navier-Stokes equations) show that separation occurs at an angle of 82° degrees from the front stagnation point on the lower and upper half of the cylinder, where the pressure gradient is actually favorable. Note however, that the theory developed above will predict separation to occur in the rear half of the cylinder. The discrepancy can be attributed to the fact that the streamwise derivatives that were neglected during the development of the theory actually become increasingly large as the separation point is approached.

The pressure distribution on the surface of the cylinder for potential flow as well as laminar flow is shown in Fig. 6.13. Owing to the loss of symmetry of the flow caused by the separation of the boundary layer in the rear half of the cylinder, there is now a net force on the cylinder in the horizontal direction. The pressure in the separated flow region is constant and much less than the pressure at the same angular location on the front half. Alternatively, the area under the pressure profile is more on the front half of the cylinder than the rear half. Consequently, the net pressure force on the cylinder is from left to right. Since this is in the same direction as the freestream, it is a drag force and is usually called *pressure drag* or *form drag*. It can thus be seen that, while the drag is zero in the potential flow, the boundary layer flow generates a drag force due to viscosity (called *friction drag*) and a drag force due to pressure force†. Mathematically, this

†Note that, in this case, the pressure drag is a consequence of the separation of the

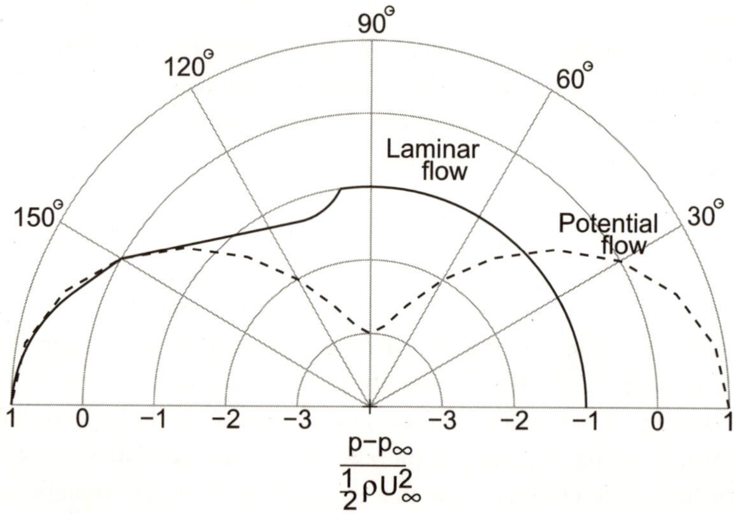

Fig. 6.13: Variation of the pressure along the surface of the cylinder for potential flow and laminar flow

can be written as

$$\mathfrak{D} = \int_{wetted\ surface} (p\,dA)_x + \int_{wetted\ surface} (\tau_{wall}\,dA)_x\,. \tag{6.38}$$

The pressure force acts normal to the wetted surface and the friction force due to the wall shear stress acts along the surface, and the components of these forces along the freestream direction (assumed to be the x-direction) in the above equation give rise to the drag force on the surface.

In the case of the cylinder, since the separated flow region is so large, the mismatch in pressure between the front and the rear side of the cylinder is also high and the pressure drag is usually much larger than the friction drag. Objects for which the pressure drag is considerably larger than the friction drag are usually called bluff bodies. Automobiles and trains are practical examples of bluff bodies. In contrast, the drag force on a streamlined body such as an aircraft is largely due to skin friction. This difference

boundary layer and the existence of the boundary layer itself can be attributed to the viscosity of the fluid. This is the reason why inviscid theory predicts not only the friction drag to be zero but the pressure drag to be zero as well.

is illustrated in Fig. 6.14 for the external flow around a flat plate for two different orientations. When the orientation is normal to the freestream direction, the drag on the plate is entirely pressure drag. On the other hand, when the plate is aligned to the freestream direction, the drag is entirely skin friction drag. The flat plate thus behaves as a bluff body in the former case and as a streamlined body in the latter case.

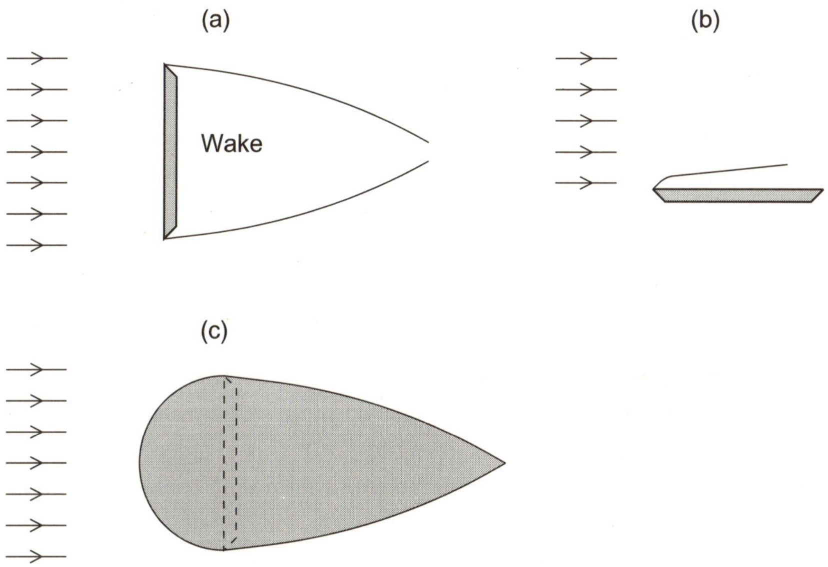

Fig. 6.14: Illustration of a flat plate oriented (a) normal to and (b) aligned with the freestream. (c) Streamlining of the flat plate to reduce pressure drag.

When the plate is oriented normal to the freestream direction, the component of velocity along that direction is reduced from the freestream value to zero on the front side. This is accompanied by an increase in the pressure in the front side. On the other hand, flow separation from the top and bottom edge of the plate results in a low pressure wake in the rear side. This combination of high pressure in the front and low pressure in the rear leads to high pressure drag in this case. This can be mitigated to a large extent by streamlining the body as shown in Fig. 6.14 (c). Streamlining the front side reduces the pressure rise there, since the x-component of velocity no longer has to be reduced to zero. Streamlining the rear side completely eliminates flow separation and the concomitant low pressure wake. However, it is important to note that the wetted surface area increases as a result of streamlining. Consequently, the reduction in pressure drag is offset by the fact that, skin friction drag, which was absent before is present now. Nevertheless, streamlining is an effective strategy for reducing drag in the case of bluff bodies†.

Separation can be delayed (or controlled) either by injecting high momentum fluid into the boundary layer (injection or blowing), or by continuously removing the slow moving fluid near the wall (suction) through slots or holes in the surface. Transition to turbulence *i.e.,* forcing the flow to become turbulent is a widely used technique and this is discussed in detail later.

6.5 Other flows governed by the boundary layer equations

In this section, we discuss two flows which are governed by the boundary layer equations Eqns. 6.1 and 6.9 and admit a self-similar solution in a manner similar to the flows discussed so far in this chapter. However, the two flows discussed in this section are different owing to the absence of a no-slip surface.

6.5.1 *Free shear layer*

Consider two streams of a fluid of density ρ, viscosity μ moving at $U_{\infty 1}$ and $U_{\infty 2}$. The streams are at the same pressure p_∞. Once the streams start co-flowing (from the origin), there is an exchange of momentum between the streams *i.e.,* the initially faster stream is decelerated while the initially slower stream is accelerated through viscous action at the interface. A shear layer of thickness (δ) grows from the origin, as shown in Fig. 6.15.

†This strategy is not possible, for instance, in the case of large signboards or billboards seen by the side of roads and highways. These are perfect examples of bluff bodies as illustrated in Fig. 6.14 (a) and can experience quite high drag forces (usually referred to as wind load). In these cases, the pressure on the front (windward) and rear (leeward) side can be equalized by punching small holes across the entire surface. Interestingly, this eliminates the pressure drag without the penalty of increased skin friction drag. Moreover, since the holes are small, there is no adverse impact on the aesthetic aspect as well.

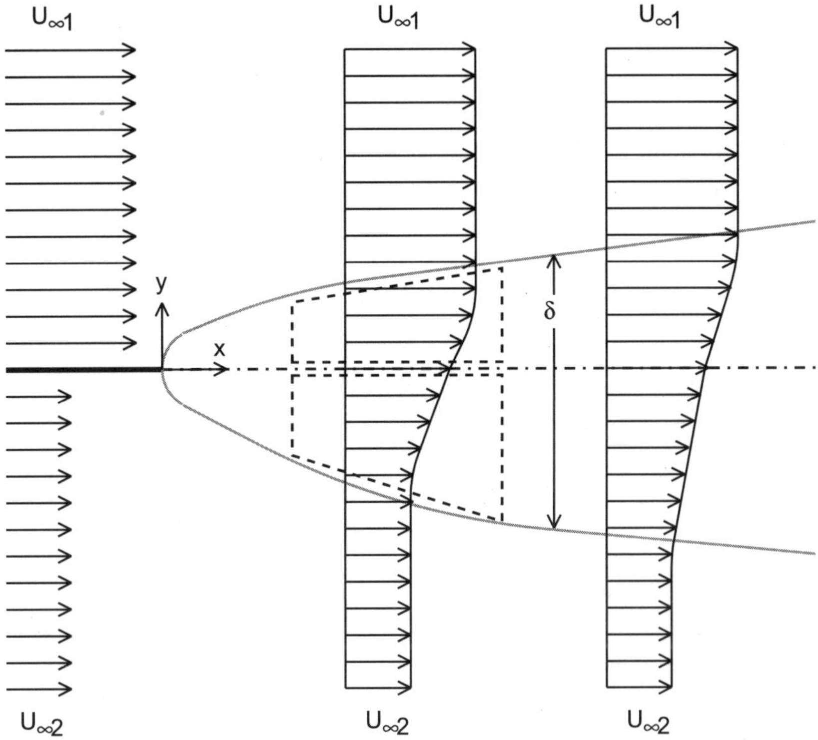

Fig. 6.15: Development of the shear layer arising from two co-flowing streams

The equations that govern the motion of the fluid are Eqns. 6.1 and 6.2 and 6.3. Boundary conditions are

$$
\begin{aligned}
u(x, y \to \infty) &= U_{\infty 1}\,, \\
u(x, y \to -\infty) &= U_{\infty 2}\,, \\
v(x, y = 0) &= 0\,,
\end{aligned}
$$

The requirement that the y-component of velocity be zero at the interface ensures that the interface remains horizontal.

An inspection of the flow inside the shear layer shows that $\partial u/\partial x$ is non-zero here but zero outside. It is clear from Eqn. 6.1 that $\partial v/\partial y$ follows the same trend. Thus, v is small but non-zero inside the shear layer, similar to what was seen earlier in a boundary layer. Consider the control volume shown in dashed line in Fig. 6.15 above the interface. Since the upper fluid is decelerated, conservation of mass within this control volume dictates that the y-component of velocity be positive on the upper face of the control volume (note that the bottom face is aligned on the interface, although it is illustrated to be slightly above in Fig. 6.15 for the sake of clarity). A similar consideration for the control volume below the interface, shows that the y-component of velocity on the bottom face of this control volume is positive, since the lower fluid is accelerated. An estimate for the magnitude of the y-component of velocity inside the shear layer can be obtained from Eqn. 6.1 as

$$V \sim \frac{U_{\infty 1}\,\delta}{x}\,.$$

It must be noted that, in the absence of a characteristic physical dimension in the streamwise direction, the axial coordinate x itself has been used as such. A scaling analysis of the x-momentum equation shows (keeping in mind that viscous action is the driving mechanism inside the shear layer) that

$$\delta \sim \sqrt{\frac{\nu\,x}{U_{\infty 1}}}\,,$$

and also that $\partial^2 u/\partial x^2 << \partial^2 u/\partial y^2$. A similar analysis of the y-momentum equation shows that $\partial p/\partial y = 0$. Since the freestream velocities $U_{\infty 1}$ and $U_{\infty 2}$ remain the same, $\partial p/\partial x = 0$ as well. Hence the governing equations simplify to

$$\frac{\partial u}{\partial x} + \frac{\partial v}{\partial y} = 0$$

and

$$u\,\frac{\partial u}{\partial x} + v\,\frac{\partial u}{\partial y} = \nu\,\frac{\partial^2 u}{\partial y^2}\,.$$

If we assume the similarity coordinate to be $\eta = y/\delta$ and that $u/U_{\infty 1}$ is a function of η only and proceed in a manner similar to what was done earlier for the flow over a flat plate with zero pressure gradient, we get

$$\psi = U_{\infty 1}\, \delta\, f(\eta)\,.$$

In deriving this expression, we have used the fact that the interface itself is a streamline, since no flow crosses it. It can further be shown that

$$\begin{aligned} u &= U_{\infty 1}\, f' \\ v &= -U_{\infty 1}\, \frac{d\delta}{dx}\, \left(f - \eta f'\right)\,. \end{aligned}$$

The governing equations reduce to

$$f''' + \frac{1}{2} f f'' = 0\,. \tag{6.39}$$

The boundary conditions given above can be rewritten as

$$\begin{aligned} f'(\eta \to \infty) &= 1\,, \\ f'(\eta \to -\infty) &= \frac{U_{\infty 2}}{U_{\infty 1}} = \lambda\,, \\ f(0) &= 0\,. \end{aligned}$$

Equation 6.39 has to be solved numerically. However, it is well known that the interface does not remain horizontal and becomes wavy even if disturbed slightly†.

6.5.2 *2D laminar jet*

Another example of a flow that is governed by the boundary layer equations is that due to a laminar jet issuing from a narrow orifice as shown in Fig. 6.16. The jet entrains the ambient air thereby exchanging momentum with

†This is called the Kelvin-Helmholtz instability.

it. Consequently, the jet spreads and decelerates, resulting in the velocity profiles shown in this figure. The width of the jet is taken as 2δ. The flow in this is also characterized by the absence of a no-slip surface and the pressure being constant everywhere. The equations that govern are Eqns. 6.1, 6.2 and 6.3. In addition, there is net conservation of momentum owing to the absence of any forces. Thus

$$\int_{-\infty}^{\infty} \rho u^2 dy = J = constant\,, \tag{6.40}$$

where ρ is the density of the jet fluid. Boundary conditions for these equations are

$$\begin{aligned} u(x, y \to \infty) &= 0\,, \\ \left.\frac{\partial u}{\partial y}\right|_{x,y=0} &= 0\,, \\ v(x, y = 0) &= 0\,. \end{aligned}$$

Owing to the symmetry of the flow field about $y = 0$, it is sufficient to obtain the solution in the upper half-plane. An order of magnitude estimate of the continuity equation reveals that

$$V \sim \frac{U_c\,\delta}{x}\,,$$

where the centerline velocity $U_c = u(x, y = 0)$ and the x-coordinate have respectively been used as the characteristic velocity and the characteristic length scale in the x-direction. A scaling analysis of the momentum equations shows that $\delta \sim \sqrt{\nu x/U_c}$, where ν is the kinematic viscosity of the jet fluid. We are thus led to Eqns. 6.1 and 6.9 as the governing equations. As before, we assume the similarity coordinate to be $\eta = y/\delta$ and that u/U_c is a function of η only. However, before we can proceed, we need to determine how the centerline velocity varies with x. This we do by rewriting Eqn. 6.40 as

$$U_c^2\,\delta \int_{-\infty}^{\infty} \left(\frac{u}{U_c}\right)^2 d\eta = \frac{J}{\rho}\,.$$

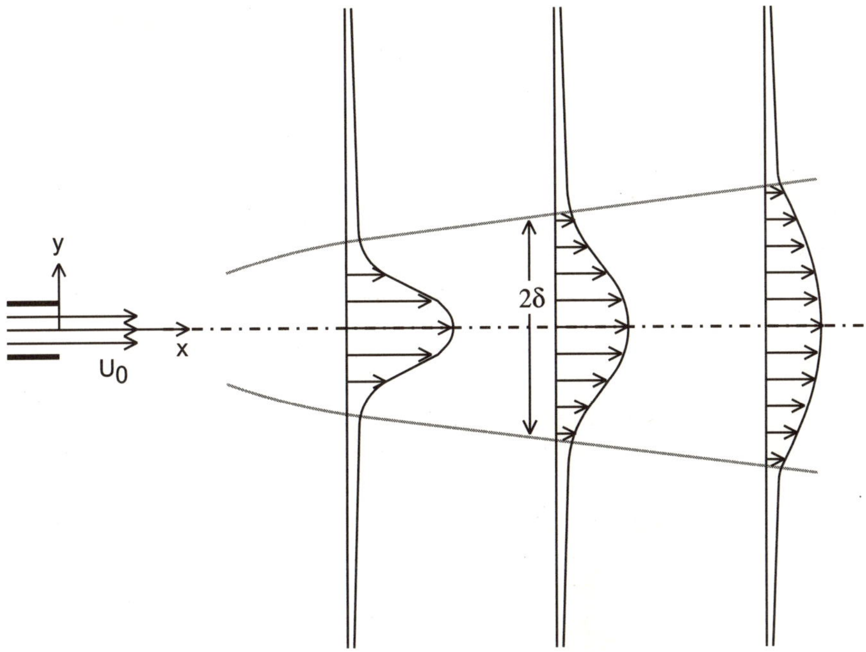

Fig. 6.16: Laminar jet issuing from a narrow slot

For the similarity hypothesis to hold, the product $U_c^2\,\delta$ in the above expression has to be a constant, as all the other quantities are constants. Thus

$$U_c^2 \sim \frac{1}{\delta} \sim \sqrt{\frac{U_c}{\nu x}} \Rightarrow U_c = \kappa \nu^{-1/3}\, x^{-1/3}\,, \tag{6.41}$$

where κ is a constant to be determined. The centerline velocity is thus seen to decay as $x^{-1/3}$. This expression also shows that the centerline velocity of an air jet, for instance, decays faster than a water jet since the kinematic viscosity of air is higher than that of water. Also, $\delta \sim \nu^{2/3}\, x^{2/3}$, which implies that the jet spreads as $x^{2/3}$.

We now proceed as before with $\eta = y/\delta$, where $\delta = \nu^{2/3}\, x^{2/3}/\kappa^{1/2}$. Since the characteristic velocity U_c is a function of x, the procedure is the same as the one leading to Eqn. 6.28. It is in fact easy to show that, in this case,

$$f''' + \alpha f f'' - \beta f'^2 = 0\,,$$

where we have taken into account the fact that the pressure gradient is zero and α and β are given in Eqn. 6.29. If we carry the differentiations through, we are finally led to

$$3f''' + f f'' + f'^2 = 0\,,$$

subject to the conditions $f(0) = f'(0) = 0$ and $f'(\eta \to \infty) = 0$. The exact solution is $f(\eta) = \sqrt{6}\tanh(\eta/\sqrt{6})$. The constant κ alone remains to be determined. This can be evaluated in terms of the prescribed value of J from Eqn. 6.40 as follows:

$$\int_{-\infty}^{\infty} U_c^2 \left(\frac{u}{U_c}\right)^2 dy = \frac{J}{\rho}$$

$$U_c^2 \int_{-\infty}^{\infty} f'^2\, \delta\, d\eta = \frac{J}{\rho}$$

$$\kappa^{3/2} \int_{-\infty}^{\infty} \mathrm{sech}^4(\eta/\sqrt{6})\, d\eta = \frac{J}{\rho}$$

$$\kappa^{3/2}\,\sqrt{6}\,\frac{4}{3} = \frac{J}{\rho}$$

$$\Rightarrow \quad \kappa = \left[\frac{3}{32}\left(\frac{J}{\rho}\right)^2\right]^{1/3}.$$

In deriving the above expression, we have used the fact that $\int_{-\infty}^{\infty} \mathrm{sech}^4(\xi)\, d\xi = 4/3$. With the exact solution available, all the quantities of interest can be evaluated. The mass flow rate at each $x = constant$ station is given as

$$\begin{aligned}
\dot{m} &= \int_{-\infty}^{\infty} \rho u dy \\
&= \rho U_c \delta \int_{-\infty}^{\infty} \left(\frac{u}{U_c}\right) d\eta \\
&= \rho \left(\frac{J}{\rho}\right)^{1/3} \left(\frac{3}{32}\right)^{1/6} \nu^{1/3} x^{1/3} \int_{-\infty}^{\infty} f' \, d\eta \\
&= 3.302\rho \left(\frac{J}{\rho}\right)^{1/3} \nu^{1/3} x^{1/3},
\end{aligned}$$

where, we have used $\int_{-\infty}^{\infty} \text{sech}^2(\xi)\, d\xi = 2$. The mass flow rate through each $x = constant$ section increases as $x^{1/3}$ as the jet entrains more air from the ambient.

Exercises

(1) Consider the flow of air over a flat at zero pressure gradient as shown in Fig. 6.5. The freestream velocity is 2 m/s and B is located at a distance of 1 m from the leading edge of the plate. Assume a quadratic profile for the velocity inside the boundary layer and that point C is located at the edge of the boundary layer. Determine the mass flow rate that leaves through the top face of the control volume and the force required to keep the plate stationary. The width of the plate may be taken as 1 m. [0.0114 kg/s, 0.009125 N]

(2) Consider a duct of square cross-section of $0.25m \times 0.25m$ and length 0.5 m. Air enters the duct with a uniform velocity of 10 m/s. If the thickness of the boundary layer is 0.04 m at the exit of the duct, determine (a) the freestream velocity at the exit and (b) the pressure drop across the duct. If the duct walls are inclined slightly so that the freestream velocity remains the same between the inlet and outlet, determine the angle of inclination. [11.207 m/s, 15.3576 Pa, 1.52°]

(3) Write a computer program to solve Eqn. 6.28. Use the program to obtain f for the flow over a 60° wedge. Derive expressions for the boundary layer thickness and wall shear stress and compare with the expressions obtained using integral analysis.

(4) Solve the Falkner-Skan equation for f for 2D plane stagnation flow. Derive expressions for the boundary layer thickness and wall shear stress.

(5) Solve the Falkner-Skan equation for f for the flow over a flat plate with a sink at the origin with $\beta = 1$. Derive expressions for the boundary layer thickness and wall shear stress.

(6) Consider the flow of an incompressible fluid over a flat plate with zero pressure gradient and wall suction as shown in the figure. Use the integral approach to determine the drag on the plate with and without suction. Assume a quadratic velocity profile for the streamwise

component of velocity. From the results that you have obtained, show whether the drag has increased or decreased as a result of the suction. $[\mathcal{D} = \frac{2}{15}\rho U_\infty^2 \delta_L + \rho U_\infty V_w L]$.

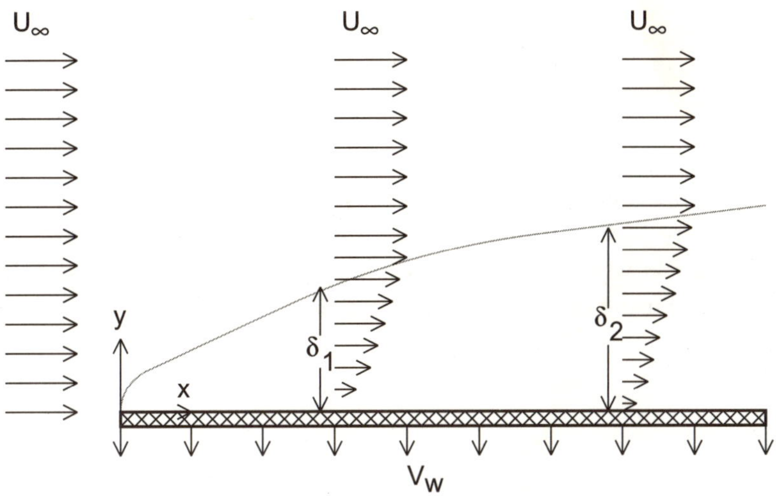

(7) Using the integral momentum analysis, derive an expression for the boundary layer thickness and the wall shear stress for 2D plane stagnation flow. Compare the results with those obtained using the Falkner-Skan equation.

(8) Consider the developing flow between parallel plates shown in the figure. At any x, let the velocity profile consist of a potential core $U(x)$ which satisfies Bernoulli's equation and a parabolic profile inside the boundary layer. The flow enters with a constant velocity U_0 and pressure P_0.
Apply the steady flow integral momentum relation and mass conservation to compute $U(x)$ and $\delta(x)$. Also determine the development length, *i.e.*, the distance from the duct entrance for the boundary layer thickness to become equal to H†.

†Sparrow, E. M., Analysis of laminar flow convection heat transfer in the entrance region of flat rectangular ducts, NACA Technical Note 3331, 1955.

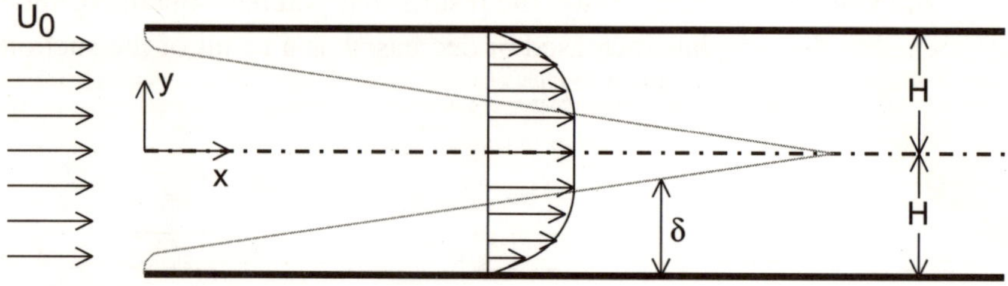

(9) Using the integral momentum analysis, derive an expression for the boundary layer thickness and the wall shear stress for the flow over a flat plate with a sink at the origin. Compare the results with those obtained using the Falkner-Skan equation.

Chapter 7

Analytical Solutions to the Incompressible Navier-Stokes Equations

In this chapter, closed form solutions to the incompressible Navier-Stokes equations are derived for a few special flow situations. The solutions can be broadly classified into two categories: those obtained under the parallel flow assumption and those obtained using the so-called "lubrication approximation". We have restricted ourselves to these solutions keeping in mind the scope of the book. The readers are urged to peruse the literature for other closed form solutions that are reported therein.

7.1 Parallel flow solutions

Parallel flows are characterized by the fact that there is a predominant flow direction. Consequently, the velocity component along this direction (streamwise component) is taken to be non-zero, while the velocity components in the other directions (cross-stream components) are taken to be zero, or, in some cases, constant. It follows immediately from the continuity equation that the streamwise component of velocity is a function only of the cross-stream coordinates and independent of the streamwise coordinate. A few parallel flow solutions are discussed next.

7.1.1 *Couette-Poiseuille flow*

Consider the flow of an incompressible fluid contained between two horizontal plates as shown in Fig. 7.1. The lower plate is held stationary while the upper plate moves along the positive x-direction at a steady speed

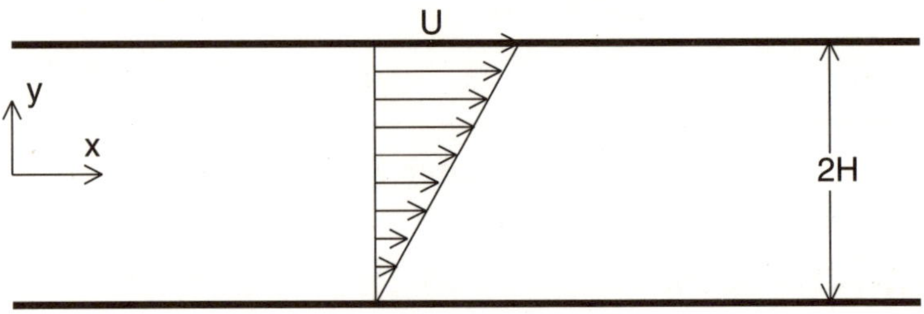

Fig. 7.1: Couette flow between two parallel plates

U. Since the flow is along the x-direction, the x-velocity component alone is non-zero. It then follows from the continuity equation, Eqn. 3.11 that $\partial u/\partial x = 0$, or, $u = u(y, z)$. If we assume the plates to be of infinite width in the z-direction, then $u = u(y)$. The x-component of the momentum equation, Eqn. 3.23 simplifies to

$$\mu \frac{d^2 u}{dy^2} = \frac{\partial p}{\partial x} \,.$$

Boundary conditions are $u = 0$ at $y = -H$ and $u = U$ at $y = H$. Since the y- and z-component of velocity are zero, $\partial p/\partial y$ and $\partial p/\partial z$ are both zero. Moreover, since $u = u(y, z)$, $\partial p/\partial x$ cannot be a function of x also and hence has to be a constant. If we set the pressure gradient to zero and solve the above equation, the solution is

$$u = U \frac{y + H}{2H} \,. \tag{7.1}$$

The velocity profile is linear and this flow is called the Couette flow. The wall shear stress on the moving plate is given as

$$\tau_{wall} = \mu \left. \frac{\partial u}{\partial y} \right|_{y=H} = \frac{\mu U}{2H} \,.$$

The power required to keep the upper plate moving is

$$\mathfrak{P} = [\tau_{wall} \times (LW)]\, U = \frac{\mu U^2 LW}{2H} \,,$$

where L is the length of the plate in contact with the fluid and W is the width perpendicular to the page.

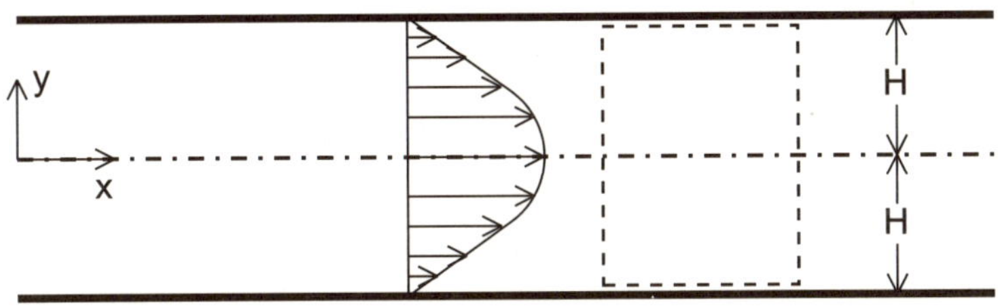

Fig. 7.2: Poiseuille flow between two parallel plates

In the case of Couette flow, the dragging action of the upper plate is responsible for driving the flow. The governing equation is homogeneous but the boundary conditions are not. Alternatively, we can keep the upper plate also stationary and drive the flow by imposing a constant pressure gradient, in which case, the governing equation would be non-homogeneous and the boundary conditions would be homogeneous. Equation 3.23 becomes

$$\mu \frac{d^2u}{dy^2} = \frac{dp}{dx} = \Lambda\,.$$

Boundary conditions are $u = 0$ at $y = \pm H$. The solution is

$$u = -\frac{\Lambda}{2\mu}\left(H^2 - y^2\right). \tag{7.2}$$

The velocity profile is parabolic and for negative values of Λ, the velocity profile looks as shown in Fig. 7.2. This flow is called the Poiseuille flow. The maximum velocity occurs at the centerline and is given as $u_{max} = -H^2\Lambda/(2\mu)$.The volumetric flow rate Q is given as

$$Q = \int_{-H}^{H} uW\,dy = -\frac{2\Lambda W H^3}{3\mu}\,,$$

where W is the width perpendicular to the page. Therefore, the pressure required to maintain a flow rate Q is given by

$$\Lambda = -\frac{3\mu Q}{2WH^3} \,. \tag{7.3}$$

The average velocity $\bar{u} = Q/(2HW) = -\Lambda H^2/(3\mu)$. It is easy to see that $\bar{u} = (2/3)u_{max}$. The wall shear stress is given as

$$\tau_{wall} = \mu \left.\frac{\partial u}{\partial y}\right|_{y=H} = \Lambda H \,.$$

Consider the control volume shown in dashed line in Fig. 7.2. There is a drag force on the upper and lower surfaces of the control volume. There is also a pressure force on the left and right faces of the control volume. There is, however, no change in the velocity of the fluid and hence no change in momentum across the control volume. This means that the drag force is balanced exactly by the pressure force. Assuming the flow to be from left to right as indicated in Fig. 7.2, *i.e.,* Λ is negative, this can be written mathematically as

$$\begin{aligned} [p\mathcal{A} - (p - |\Lambda|L)\mathcal{A}] &= |\tau_{wall}|\mathcal{P}L \\ \Lambda| &= \frac{\mathcal{P}}{\mathcal{A}}\,|\tau_{wall}| \,, \end{aligned} \tag{7.4}$$

where, L is the length of the control volume, $\mathcal{A}$ is the cross-sectional area and $\mathcal{P}$ is the wetted perimeter. Since the flow is assumed to be two-dimensional, there are no confining walls except those on the top and bottom and hence the shear stress is exerted only on these walls. This has to be taken into account while calculating $\mathcal{P}$. Thus,

$$|\Lambda| = \frac{2W}{2H \times W} = \frac{1}{H}\,|\tau_{wall}| \,.$$

Substitution of the expression for wall shear stress from above shows that this expression is satisfied identically. Physically, this means that the pressure gradient that drives the flow should exactly match the pressure drop due to friction at the wall, so that the flow neither accelerates nor decelerates.

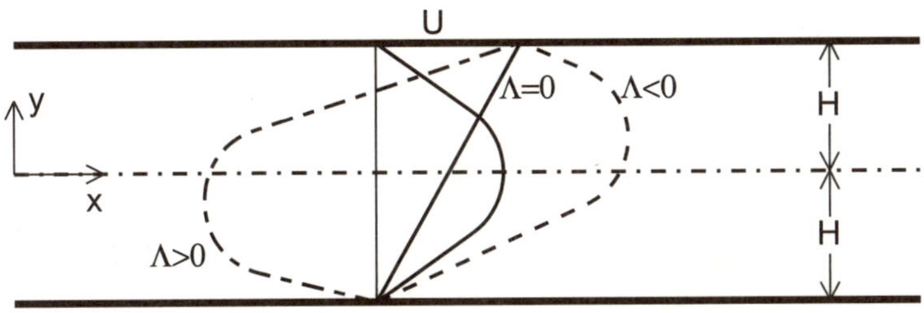

Fig. 7.3: Couette Poiseuille flow between two parallel plates

If we allow the upper plate to move as before and impose a pressure gradient as well, the resulting flow is usually called a Couette Poiseuille flow. The resulting velocity profile can be written as

$$u = U\frac{y+H}{2H} - \frac{\Lambda}{2\mu}\left(H^2 - y^2\right). \tag{7.5}$$

This can be written in dimensionless form as

$$\frac{u}{U} = \frac{1}{2}\left(1 + \frac{y}{H}\right) - \frac{H^2\Lambda}{2\mu U}\left(1 - \frac{y^2}{H^2}\right).$$

The actual shape of the velocity profile is determined by the dimensionless pressure gradient $H^2\Lambda/(2\mu U)$. If $\Lambda = 0$, then we recover the Couette flow solution. If $\Lambda > 0$, then the imposed pressure gradient opposes the flow caused by the upper plate and the velocity profile is as shown in Fig. 7.3. Note that the velocity is negative in part of the channel in this case. If $\Lambda < 0$, then the imposed pressure gradient aids the flow caused by the upper plate and the resulting velocity profile is as shown in Fig. 7.3. The exact value for Λ has be to be determined on a case by case basis.

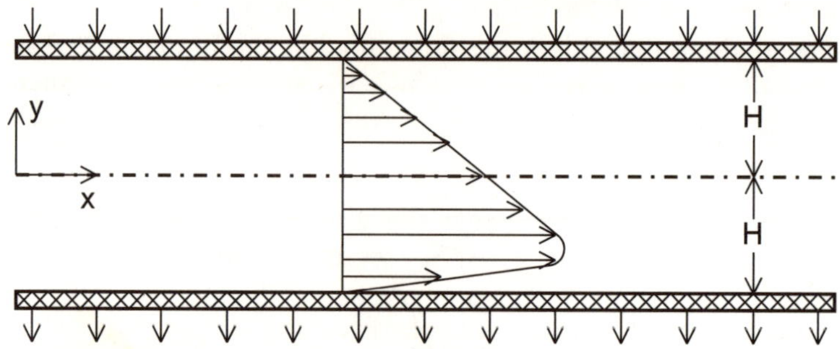

Fig. 7.4: Poiseuille flow between two parallel, porous plates

As mentioned earlier, the parallel flow assumption can be used even when the velocity component is constant but non-zero along a coordinate direction besides the flow direction. The Poiseuille flow solution in a channel derived above can be extended to the case when there is a cross-flow along the y-direction due to the walls being porous (Fig. 7.4).

The flow is parallel to the x-axis and $\partial u/\partial x = 0$ from the continuity equation. Thus, $u = u(y)$ and $v = -V$ (since the cross-flow is along the negative y-direction). Equation 3.23 becomes

$$\nu \frac{d^2u}{dy^2} + V\frac{du}{dy} = -\frac{1}{\rho}\frac{dp}{dx} = \frac{\Lambda}{\rho}\,.$$

Boundary conditions are $u = 0$ at $y = \pm H$. The solution is given as

$$u = \frac{\Lambda H^2}{\mu Re_V}\left[1 + \frac{y}{H} - 2\left(\frac{1 - e^{-Re_V(1+y/H)}}{1 - e^{-2Re_V}}\right)\right], \qquad (7.6)$$

where, $Re_V = VH/\nu$ is the cross-stream Reynolds number. The velocity profile is shown in Fig. 7.4. A boundary layer can clearly be discerned near the lower plate at $y = -H$. Away from this layer, the exponential term in Eqn. 7.6 is essentially a constant and the velocity profile is linear. This layer becomes thinner with increasing Re_V. As $Re_V \to 0$, the Poiseuille solution (Eqn. 7.2) is recovered†.

†This is not quite obvious from Eqn. 7.6. However, a careful application of L'Hôspital's rule will indeed show that Eqn. 7.2 is recovered.

As already mentioned, the streamwise velocity u need not be a function of y alone; it can be a function of (y, z). This happens, for instance, when we consider the flow through a channel that is rectangular in cross section *i.e.*, $-H \leq y \leq H$ and $-W \leq z \leq W$. The x-momentum equation, Eqn. 3.23, with the parallel flow assumption simplifies to

$$\mu\left(\frac{d^2u}{dy^2} + \frac{d^2u}{dz^2}\right) = \frac{\partial p}{\partial x} = \Lambda\,. \tag{7.7}$$

Boundary conditions are $u = 0$ at $y = \pm H$ and $z = \pm W$. It can easily be verified that the equation above cannot be solved by using separation of variables. We assume a solution of the form

$$u(y, z) = \sum_{m=1}^{\infty}\sum_{n=1}^{\infty} u_{mn} \sin\left(m\pi\frac{y+H}{2H}\right) \sin\left(n\pi\frac{z+W}{2W}\right)\,. \tag{7.8}$$

Note that the assumed form for u satisfies the boundary conditions identically. The coefficients u_{mn} are yet to be determined. If we substitute Eqn. 7.8 into the left hand side of Eqn. 7.7 we obtain,

$$-\frac{\pi^2}{4}\sum_{m=1}^{\infty}\sum_{n=1}^{\infty} u_{mn}\left(\frac{m^2}{H^2} + \frac{n^2}{W^2}\right) \sin\left(m\pi\frac{y+H}{2H}\right) \sin\left(n\pi\frac{z+W}{2W}\right)$$

$$= \frac{\Lambda}{\mu}\,. \tag{7.9}$$

The coefficients u_{mn} can be evaluated now by exploiting the fact that

$$\int_{-H}^{H} \sin\left(m\pi\frac{y+H}{2H}\right) \sin\left(i\pi\frac{y+H}{2H}\right) dy = \begin{cases} 0, & i \neq m \\ H, & i = m \end{cases}$$

and

$$\int_{-W}^{W} \sin\left(n\pi\frac{z+W}{2W}\right) \sin\left(j\pi\frac{z+W}{2W}\right) dz = \begin{cases} 0, & j \neq n \\ W, & j = n \end{cases}\,.$$

Furthermore,

$$\int_{-H}^{H} \sin\left(i\pi\frac{y+H}{2H}\right)\,dy = \begin{cases} 0, & \text{i even} \\ \frac{4H}{i\pi}, & \text{i odd} \end{cases}$$

and

$$\int_{-W}^{W} \sin\left(j\pi\frac{z+W}{2W}\right)\,dz = \begin{cases} 0, & \text{j even} \\ \frac{4W}{j\pi}, & \text{j odd} \end{cases} .$$

If we multiply both sides of Eqn. 7.9 by $\sin\left(i\pi\frac{y+H}{2H}\right)\ \sin\left(j\pi\frac{z+W}{2W}\right)$ and integrate, we get

$$u_{mn} = -\frac{\Lambda}{\mu}\frac{64}{\pi^4}\frac{1}{mn}\frac{1}{\left(\frac{m^2}{H^2}+\frac{n^2}{W^2}\right)} \qquad \text{m,n odd.}$$

With the velocity known, the flow rate Q can be calculated as

$$\begin{aligned} Q &= \int_{-W}^{W}\int_{-H}^{H} u\,dy\,dz \\ &= -\frac{\Lambda}{\mu}\frac{256}{\pi^6}4HW\sum_{m=1}^{\infty}\sum_{n=1}^{\infty}\frac{1}{m^2n^2}\frac{1}{\left(\frac{m^2}{H^2}+\frac{n^2}{W^2}\right)}, \quad \text{m,n odd.} \end{aligned} \qquad (7.10)$$

The average velocity $\bar{u} = Q/4HW$ for several values of W/H are given in Table. 7.1.

Table 7.1: Average velocity for laminar flow in a rectangular duct at different aspect ratios

W/H	1	2	4	8	16	∞
$\frac{\bar{u}}{-\Lambda H^2/3\mu}$	0.4217	0.6860	0.8424	0.9212	0.9606	1

It is, of course, comforting to note that as $W/H \to \infty$, the expression for $\bar{u}$ is identical to the one obtained earlier for the flow field sketched in Fig. 7.2, as it should be.

7.1.2 *Hagen-Poiseuille flow in a pipe*

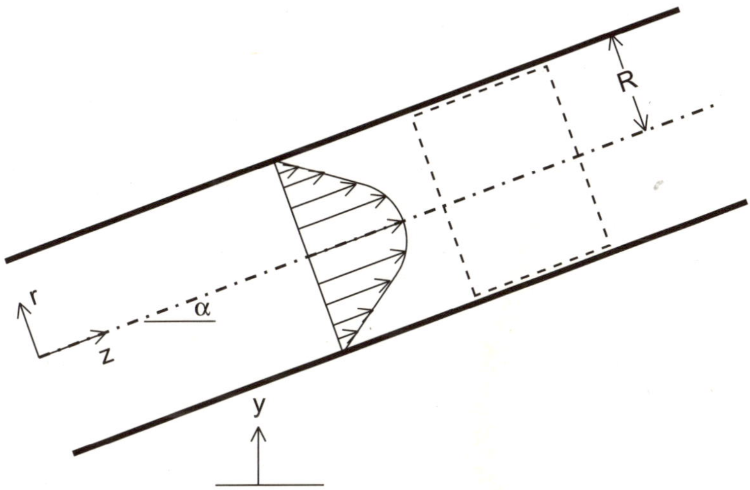

Fig. 7.5: Hagen Poiseuille flow in a pipe

We now turn to laminar flow through a pipe which is also a parallel flow†. The flow direction is along the length of the pipe. Hence, the axial component of velocity, w, alone is non-zero. From the continuity equation, Eqn. A.18, it follows that $w = w(r)$. The z-component of the momentum equation, Eqn. A.21, reduces to

$$\mu\left(\frac{d^2w}{dr^2} + \frac{1}{r}\frac{dw}{dr}\right) = \rho g \sin\alpha + \frac{dp}{dz} = \rho g \sin\alpha + \Lambda\,, \qquad (7.11)$$

where we have set $F_z = -\rho g \sin\alpha$. The pressure gradient has been set equal to a constant Λ, based on the same argument as the one made earlier

†Except near the entrance of the pipe, where the growing boundary layers on the wall render the flow two dimensional. For this reason, the parallel flow solution is usually said to be fully developed. This applies to the Poiseuille flow solution discussed earlier as well.

for the flow through a channel. Boundary conditions are that $w = 0$ at $r = R$, where R is the radius of the pipe and that the velocity should be finite on the axis ($r = 0$). The equation above can be rewritten as

$$\frac{1}{r}\frac{d}{dr}\left(r\frac{dw}{dr}\right) = \frac{\rho g \sin\alpha + \Lambda}{\mu}.$$

Here, α is the inclination of the pipe to the horizontal. If we integrate this once, we get

$$r\frac{dw}{dr} = \frac{\rho g \sin\alpha + \Lambda}{\mu}\frac{r^2}{2} + C_1,$$

where C_1 is a constant. If we integrate this once more, we get

$$w = \frac{\rho g \sin\alpha + \Lambda}{\mu}\frac{r^2}{4} + C_1 \ln r + C_2,$$

where, C_2 is a constant. Since w has to be finite at $r = 0$, the constant $C_1 = 0$. Hence,

$$w = -\frac{\rho g \sin\alpha + \Lambda}{4\mu}(R^2 - r^2). \tag{7.12}$$

The velocity profile is a paraboloid as shown in Fig. 7.5 and the flow is called the Hagen-Poiseuille flow in a pipe. Note that, the pressure gradient Λ is negative for the flow direction shown in Fig. 7.5. Furthermore, for w to be positive, $|\Lambda| > \rho g \sin\alpha$ when the pipe slopes upwards. The maximum velocity occurs at the centerline and is given as $w_{max} = -R^2(\rho g + \Lambda)/(4\mu)$. The volumetric flow rate is

$$Q = \int_0^R w \times 2\pi r\, dr = -\frac{\pi(\rho g \sin\alpha + \Lambda)}{8\mu}R^4.$$

The average velocity $\bar{w} = Q/(\pi R^2) = -(\rho g \sin\alpha + \Lambda)R^2/(8\mu)$. It is easy to see that $\bar{w} = w_{max}/2$. Although the velocity profiles in Figs. 7.5 and

7.2 bear a resemblance, the ratio of the average to the maximum velocity is different between an axisymmetric pipe flow and the flow between parallel plates.

Let us consider two points 1 and 2 located a distance L apart along the axis of the pipe, with point 1 being upstream. The change in pressure between these two points is due to (a) a change in the elevation and (b) friction. The latter always causes the pressure to decrease, resulting in the so-called head loss. When the pipe slopes upwards along the direction of the flow, there is a reduction in the hydrostatic component of the pressure along the length of the pipe. On the other hand, when the pipe slopes downwards, the hydrostatic component increases along the flow direction. The pressure loss due to friction equals the pressure change, Δp (= $p_2 - p_1$) minus the change in the hydrostatic component, $-\rho g \Delta y$ where $\Delta y = y_2 - y_1$. Thus, the head loss due to friction h_f, between these two points can be written as

$$h_f = -\frac{\rho g \Delta y + \Delta p}{\rho g} \tag{7.13}$$

$$= \frac{1}{\rho g}\frac{8\mu Q L}{\pi R^4} \tag{7.14}$$

$$= \frac{1}{\rho g}\frac{8\mu \bar{w} L}{R^2}\,. \tag{7.15}$$

Equation 7.14 shows that the head loss due to friction depends on the viscosity, flow rate, length and pipe radius and is independent of the inclination of the pipe. It is customary to write the head loss as

$$h_f = \frac{f L \bar{w}^2}{2gD}\,, \tag{7.16}$$

where $D = 2R$ is the pipe diameter and f is the friction factor†. Equation

†Engineering calculations involving pipes and pipe networks often combine Eqn. 7.16 with Eqn. 5.14 to arrive at

$$\frac{p_1}{\rho g} + z_1 + \frac{\bar{w}_1^2}{2g} = \frac{p_2}{\rho g} + z_2 + \frac{\bar{w}_2^2}{2g} + h_f$$

7.16 is called the Darcy-Weisbach law. If we compare Eqns. 7.15 and 7.16, it follows that the friction factor for laminar flow is

$$f = \frac{1}{\rho}\frac{8\mu}{R^2}\frac{2D}{\overline{w}} = \frac{64\mu}{\rho\overline{w}D} = \frac{64}{Re_D}, \tag{7.17}$$

where $Re_D = \rho\overline{w}D/\mu$ is the Reynolds number based on the average velocity and the pipe diameter.

The wall shear stress is given as

$$\tau_{wall} = \mu \left.\frac{\partial w}{\partial r}\right|_{r=R} = \frac{(\rho g \sin\alpha + \Lambda)R}{2} = \frac{4\mu Q}{\pi R^3}.$$

An integral analysis using the control volume shown in Fig. 7.5 shows that, in this case also, the pressure drop Δp that drives the flow should exactly match the pressure drop due to friction at the wall and the change in pressure due to the change in gravity head, so that the flow neither accelerates or decelerates†. Equation 7.4 thus holds true in this case also and can be written as

$$\frac{\rho g h_f}{L} = \frac{\mathcal{P}}{\mathcal{A}}\tau_{wall}.$$

With $\mathcal{P} = 2\pi R$ and $\mathcal{A} = \pi R^2$, and with h_f given by Eqn. 7.16, the above equation becomes

$$\frac{\tau_{wall}}{\rho\overline{w}^2} = \frac{f}{8}. \tag{7.18}$$

It is thus clear that the friction factor f is nothing but a dimensionless shear stress. The pumping power required to maintain a volume flow rate Q is the sum of the power required to overcome head loss and the power required

---

Strictly speaking, Bernoulli's equation is not applicable in such cases since the flow is not inviscid. However, it may be considered as an engineering approximation and the simplicity of the above equation and the usefulness of the results has led to its extensive use.

†In other words, a constant flow rate is maintained.

to achieve the change in gravity head. Thus

$$\mathfrak{P} = \rho g \Delta y \, Q + [\tau_{wall}(2\pi RL)] \, \bar{w}$$

$$= \rho g \Delta y \, Q + \frac{128\mu Q^2 L}{\pi D^4} \qquad (7.19)$$

$$= \rho g (\Delta y + h_f) \, Q \, . \qquad (7.20)$$

Equations 7.19 and 7.14 show that the pumping power and the head loss are both inversely proportional to D^4. If the diameter of the pipe is doubled, the power and the head loss both decrease by a factor of 16. Of course, the weight of the pipe and hence the cost increases approximately by a factor of 4 (with thickness remaining the same).

It is of interest to note that Eqn. 7.14 is usually used to determine the viscosity of a fluid in fluid mechanics laboratory experiments.

It must also be clear from the development that the velocity profile, Eqn. 7.12 is not affected by the roughness of the pipe. It will be shown later that roughness plays an important role in the case of turbulent pipe flow.

Example 7.1. The Trans-Alaska pipeline (48 inches diameter) transports 754000 barrels of crude oil (specific gravity 0.93, $\mu = 24 \times 10^{-3} kg/(m.s)$) per day over a distance of 800 miles. Determine the pumping power required assuming laminar flow.

Solution. Given

$$Q = 754000 \frac{barrels}{day} \times 42 \frac{gallons}{barrel} \times 3.785 \frac{litres}{gallon} \times \frac{1}{1000} \frac{m^3}{litre}$$

$$\times \frac{1}{24 \times 3600} \frac{day}{seconds} = 1.387 m^3/s.$$

The average velocity $\bar{w} = 4Q/(\pi D^2) = 1.19 \ m/s$. The Reynolds number based on the pipe diameter Re_D comes out to be 56128. It is well known that the flow in a pipe becomes turbulent beyond $Re_D \approx 2300$, and so the assumption of laminar flow is highly unrealistic in this case. However, we proceed to calculate $f = 0.00114$ from Eqn. 7.17. The pressure drop can be calculated from Eqn. 7.16,

$$\Delta p = \rho g h_f = \frac{\rho f L \bar{w}^2}{2D} = 0.788 \times 10^6 \ N/m^2 \,.$$

The pumping power is thus

$$\mathfrak{P} = \Lambda Q L = \Delta p Q = 1.1 \ MW \ \square$$

It is easy to obtain a parallel flow solution for the case when the pipe has an elliptic cross-section. Let us assume that the cross-section is an ellipse described by the equation

$$\frac{z^2}{a^2} + \frac{y^2}{b^2} - 1 = 0 \,. \tag{7.21}$$

It is possible to write the governing equation using elliptical coordinates and proceed in a manner similar to what was done for pipes with a circular cross-section. However, it is easier to work with the equation in cartesian coordinates. Accordingly, the x-momentum equation in this case becomes,

$$\mu \left(\frac{\partial^2 u}{\partial y^2} + \frac{\partial^2 u}{\partial z^2} \right) = \frac{dp}{dx} = \Lambda \,.$$

Boundary condition is that $u = 0$ on the wall of the pipe *i.e.,* along the curve whose equation is given by Eqn. 7.21. This suggests that we look for a solution of the form

$$u(y, z) = C \left(\frac{z^2}{a^2} + \frac{y^2}{b^2} - 1 \right) ,$$

where C is a constant[†]. Substitution of this guessed profile into the governing equation leads to the following solution:

$$u(y,z) = \frac{1}{2\mu}\frac{\Lambda a^2 b^2}{a^2+b^2}\left(\frac{z^2}{a^2}+\frac{y^2}{b^2}-1\right), \qquad (7.22)$$

where $\Lambda < 0$ as before. If we set $a = b = R$ and recognize the fact that $z^2 + y^2 = r^2$, then the solution corresponding to a circular pipe, namely, Eqn. 7.12, is identically recovered. The volumetric flow rate is

$$\begin{aligned} Q &= 4\int_0^b \int_0^{a\sqrt{1-\frac{y^2}{b^2}}} u(y,z)\,dz\,dy \\ &= -\frac{\pi\Lambda}{4\mu}\frac{a^3 b^3}{a^2+b^2}. \end{aligned}$$

The average velocity

$$\bar{w} = \frac{Q}{\pi ab} = -\frac{\Lambda}{4\mu}\frac{a^2 b^2}{a^2+b^2}.$$

The head loss in this case can be written as

$$\begin{aligned} \text{Head loss} &= -\frac{\Lambda L}{\rho g} \\ &= \frac{1}{\rho g}\frac{4\mu Q L}{\pi}\frac{a^2+b^2}{a^3 b^3}. \end{aligned} \qquad (7.23)$$

Although it is possible to obtain a closed form solution for this case, in general, for pipes of non-circular cross-section, it is customary in engineering practice to use the concept of a hydraulic diameter. The

[†]Note that a similar guess for the flow in a pipe of rectangular cross-section, namely,

$$u(y,z) = C\left(1-\frac{y^2}{H^2}\right)\left(1-\frac{z^2}{W^2}\right)$$

does not work.

hydraulic diameter D_h is defined as $D_h = 4\mathcal{A}/\mathcal{P}$, where $\mathcal{A}$ is the cross-sectional area and $\mathcal{P}$ is the wetted perimeter. The hydraulic diameter is then used in place of the diameter D in Eqn. 7.17 and in other equations. Note that the hydraulic diameter is equal to the diameter D, in the case of a pipe with circular cross-section. It is important to understand that the concept of hydraulic diameter is only an engineering approximation.

Example 7.2. For laminar flow through a pipe of elliptical cross-section for which $a = 2b$, determine the error in using the hydraulic diameter to calculate the pressure drop when compared to Eqn. 7.23.

Solution. The pressure drop per unit length from Eqn. 7.23 is

$$\Lambda = -\frac{5\mu Q}{2\pi}\frac{1}{b^4}.$$

The cross-sectional area of the pipe is $\pi ab = 2\pi b^2$. The perimeter of the ellipse can be calculated from Ramanujan's formula as $\mathcal{P} = \pi[3(a + b) - \sqrt{(3a+b)(a+3b)}] = 3.084\pi b$. Therefore, the hydraulic diameter $D_h = 4\mathcal{A}/\mathcal{P} = 2.594\, b$. The pressure drop per unit length from Eqn. 7.16 is

$$\Lambda = -\frac{8\mu Q}{\pi}\frac{16}{D_h^4} = -\frac{2.827\,\mu Q}{\pi}\frac{1}{b^4}.$$

The error in using the hydraulic diameter is thus 13.08%. □

Example 7.3. For Poiseuille flow between two parallel plates, determine the error in using the hydraulic diameter to calculate the friction factor when compared to Eqn. 7.3.

Solution. The cross-sectional area of the channel is $\mathcal{A} = 2HW$, where W is the width perpendicular to the page. The wetted perimeter is $\mathcal{P} = 2W$, since top and bottom surfaces alone are wetted and there are no side walls. The hydraulic diameter $D_h = 4\mathcal{A}/\mathcal{P} = 4H$. The friction factor from Eqn. 7.17 is

$$f = \frac{64}{\mathcal{R}e_{D_h}} = \frac{16\mu}{\rho\bar{u}H}.$$

The head loss from Eqn. 7.3 is

$$\begin{aligned}\text{Head loss} &= -\frac{\Lambda L}{\rho g} \\ &= \frac{1}{\rho g}\frac{3\mu\bar{u}L}{H^2} \\ &= \frac{fL\bar{u}^2}{2g(2H)}\,.\end{aligned}$$

The friction factor is then

$$f = \frac{12\mu}{\rho\bar{u}H}\,.$$

The error in using the hydraulic diameter is thus 33.33%. $\square$

7.1.3 *Flow between concentric rotating cylinders*

Consider the flow of an incompressible fluid contained between two concentric rotating cylinders. It is easy to see that the flow is along the θ direction and that $v = v(r)$. The radial and azimuthal momentum equations (Eqns. A.19 and A.20) reduce to

$$\begin{aligned}\rho\frac{v^2}{r} &= \frac{dp}{dr} \\ r^2\frac{d^2v}{dr^2} + r\frac{dv}{dr} - v &= 0\,.\end{aligned}$$

with boundary conditions $v = R_1\Omega_1$ at $r = R_1$ and $v = R_2\Omega_2$ at $r = R_2$ ($R_2 > R_1$). The radial pressure gradient arises due to the centrifugal force. The second equation can be written as

$$\frac{d}{dr}\left(\frac{dv}{dr} + \frac{v}{r}\right) = 0\,.$$

If we integrate this, we get

$$\frac{dv}{dr} + \frac{v}{r} = C_1 ,$$

where C_1 is a constant. This can be rewritten as

$$\frac{1}{r}\frac{d}{dr}(rv) = C_1 .$$

Integrating this and applying the boundary conditions leads to

$$v = \frac{R_1\Omega_1}{1-(R_1/R_2)^2}\left[\frac{r}{R_1}\left(\frac{\Omega_2}{\Omega_1} - \frac{R_1^2}{R_2^2}\right) - \frac{R_1}{r}\left(\frac{\Omega_2}{\Omega_1} - 1\right)\right] . \tag{7.24}$$

It is noteworthy that, contrary to all the velocity profiles obtained earlier, Eqn. 7.24 does not contain the viscosity. We can now look at some special cases:

- $\Omega_1 = 0$. In this case,

$$v = \frac{R_1\Omega_2}{1-(R_1/R_2)^2}\left(\frac{r}{R_1} - \frac{R_1}{r}\right) .$$

- $\Omega_2 = 0$. In this case,

$$v = \frac{R_1\Omega_1}{1-(R_1/R_2)^2}\left(-\frac{r}{R_1}\frac{R_1^2}{R_2^2} + \frac{R_1}{r}\right) .$$

- $\Omega_1 = \Omega_2$. In this case, the flow corresponds to that of a solid body rotation.

$$v = r\Omega_1 .$$

- $R_1 \to 0$. This corresponds to solid body rotation $v = r\Omega_2$ (Fig. 5.5).

- $R_2 \to \infty$ and $\Omega_2 \to 0$. This corresponds to the case of a cylinder rotating in an infinite body of fluid. The solution for this case is

$$v = \frac{R_1^2 \Omega_1}{r} .$$

This solution also corresponds to the flow induced by vortex of strength $\Gamma = 2\pi R_1^2 \Omega_1$ in an *inviscid* fluid (Eqn. 5.20).

The wall shear stress is given as

$$\tau_{wall} = \mu\, r \frac{\partial}{\partial r}\left(\frac{v}{r}\right)\Bigg|_{wall} .$$

The torque exerted by the inner and outer cylinders are equal and this can be calculated as

$$\mathfrak{T} = \frac{4\pi\mu R_1^2 H \Omega_1}{1-(R_1/R_2)^2}\left(\frac{\Omega_2}{\Omega_1} - 1\right) , \tag{7.25}$$

where H is the height of the cylinder. If one of the cylinders is stationary, then the torque calculated from the above expression is the torque required to keep that cylinder stationary. One of the special cases mentioned above, namely, the arrangement with the inner cylinder rotating and outer cylinder stationary, is utilized in the Brookfield viscometer to measure viscosity through the above expression.

7.1.4 *Flow in convergent and divergent channels*

The flow in convergent and divergent channels (Fig. 7.6) is another example of a parallel flow where an exact solution to the incompressible Navier-Stokes equations is possible. The flow can be said to be parallel since the radial component of the velocity alone is non-zero. The streamlines are, however, radial lines and not parallel to each other. The flow field in both the cases is symmetric about the channel centerline. Furthermore, it must be recalled that the potential flow (outer) and the boundary layer (inner) solution discussed in section 6.3.1.3 is the composite solution to the flow in one half of the convergent channel (Figs.

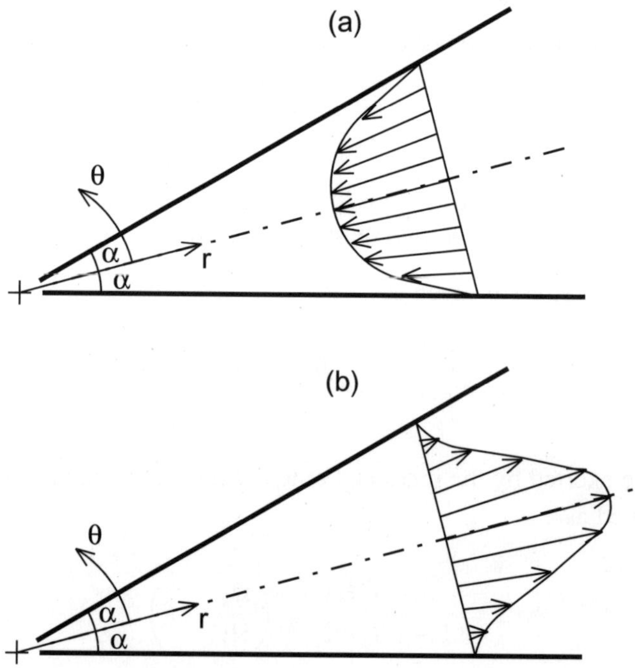

Fig. 7.6: Flow in (a) convergent and (b) divergent channels

6.9 and 7.6). It was also remarked in section 6.3.1.3 that a corresponding solution for the flow in a diverging channel could not be obtained within the ambit of boundary layer theory.

The governing equations in this case reduce to,

$$\frac{1}{r}\frac{\partial}{\partial r}(ru) = 0\,,$$

$$u\frac{\partial u}{\partial r} = -\frac{1}{\rho}\frac{\partial p}{\partial r} + \nu\left[\frac{1}{r}\frac{\partial}{\partial r}\left(r\frac{\partial u}{\partial r}\right) - \frac{u}{r^2} + \frac{1}{r^2}\frac{\partial^2 u}{\partial \theta^2}\right],$$

$$0 = -\frac{1}{\rho r}\frac{\partial p}{\partial \theta} + \nu\left(\frac{2}{r^2}\frac{\partial u}{\partial \theta}\right).$$

Boundary conditions are $u = 0$ at $\theta = \pm\alpha$ and $\partial u/\partial\theta = 0$ at $\theta = 0$.

Upon integrating the continuity equation above, we get $u = F(\theta)/r$, where F is a function in θ that is to be determined. It is in fact more convenient to write $u = \nu F(\theta)/r$. If we substitute this expression for u into the remaining two equations, we get

$$-\frac{\nu^2}{r^3}F^2 = -\frac{1}{\rho}\frac{\partial p}{\partial r} + \frac{\nu^2}{r^3}F'' ,$$
$$0 = -\frac{1}{\rho r}\frac{\partial p}{\partial \theta} + \frac{2\nu^2}{r^3}F' .$$

The pressure gradient term can be eliminated by cross differentiation and this leads to

$$F''' + 4F' + 2FF' = 0 ,$$

with boundary conditions $F(\pm\alpha) = F'(0) = 0$.

Both numerical and analytical solutions for this equation have been given by Jeffery (1915), Hamel (1916) and Rosenhead (1940). The flow rate through the channel is given as

$$|Q| = \left|\int_{-\alpha}^{\alpha} u \times r d\theta\right| = \nu \left|\int_{-\alpha}^{\alpha} F d\theta\right| .$$

The solution can be parameterized by a Reynolds number $Re = |Q|/\nu$. Velocity profiles are sketched in Fig. 7.7 for different Re.

In the case of a convergent channel, a boundary layer near both the walls can clearly be discerned from the velocity profiles at higher Re. Of course, as mentioned earlier, at such Re, it is possible to obtain a composite solution of an outer potential flow and an inner boundary layer flow.

In the case of a divergent channel, the pressure gradient is adverse and the flow decelerates continuously. Hence, at low flow rates (or, equivalently low Re), the velocity near the walls becomes negative as seen in Fig. 7.7,

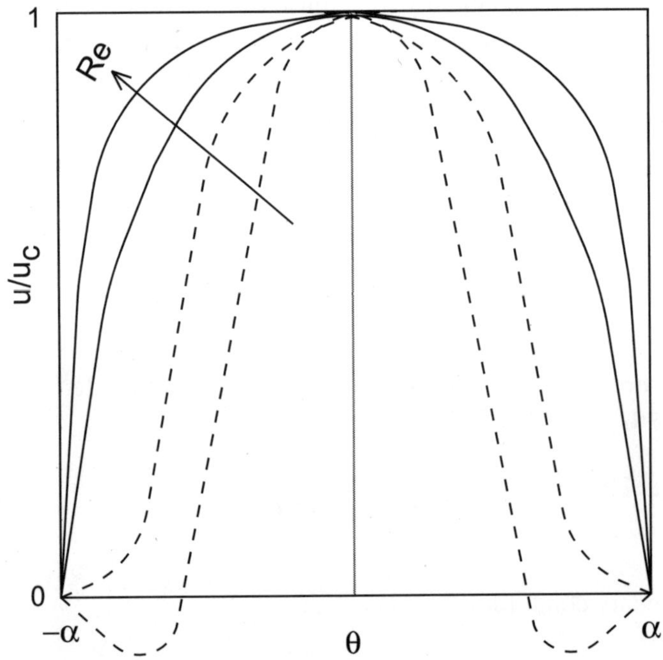

Fig. 7.7: Velocity profiles at different Reynolds numbers for the flow in convergent (solid line) and divergent (dashed line) channels

suggestive of separation. The velocity profiles at higher values of Re do not exhibit any boundary layer, which explains why a solution to this case cannot be obtained from the boundary layer equations.

7.2 Creeping flow solutions

Creeping flows are flows for which the Reynolds number $Re = Uh/\nu$ is very small. Here, h is a characteristic dimension. A good example of a creeping flow is the flow of volcanic lava. The Reynolds number in these flows are small, in general, due to the velocity (characterized by U) being very small and the viscosity ν, being very large. In this section, we will discuss the creeping flow that occurs in the narrow gap between a sliding surface and a stationary surface in a bearing. Here, the Reynolds number is small, on account of the fact that the gap width h is very small.

Let a viscous fluid occupy the small gap h between two surfaces located at $y = 0$ and $y = h$. Let U be a characteristic velocity and L, a characteristic dimension along the x-direction, such that $h \ll L$. The equations that govern the flow are the continuity equation 6.1 and the x-momentum equation 6.2 and the y-momentum equation 6.3. The sizes of the two terms in Eqn. 6.1 are

$$\frac{U}{L} \quad \frac{V}{h},$$

where, V is an estimate of the magnitude of the y-component of velocity inside the gap region. Since the x-component of velocity is expected to vary with x and y, $\partial u/\partial x$ is not zero and so these two terms are of the same size. Thus

$$V \sim \frac{Uh}{L}.$$

The sizes of the terms in Eqn. 6.2 are

$$\frac{U^2}{L} \quad \frac{VU}{h} \quad \frac{P}{\rho L} \quad \frac{\nu U}{L^2} \quad \frac{\nu U}{h^2},$$

where P is an estimate of the pressure in the gap region. If we substitute for V, we get

$$\frac{U^2}{L} \quad \frac{U^2}{L} \quad \frac{P}{\rho L} \quad \frac{\nu U}{L^2} \quad \frac{\nu U}{h^2}.$$

If we compare the two viscous terms, it is clear that the term with the h^2 in the denominator is much larger than the other one, since $h \ll L$. Hence the term $\nu \partial^2 u/\partial x^2$ can be dropped. The sizes of the remaining terms are

$$\frac{U^2}{L} \quad \frac{U^2}{L} \quad \frac{P}{\rho L} \quad \frac{\nu U}{h^2}.$$

The two convective terms are of the same size and if we compare this with the size of the remaining viscous term we get

$$\frac{U^2}{L} \qquad \frac{\nu U}{h^2} \,.$$

Or

$$\frac{Uh}{\nu}\frac{h}{L} \qquad 1\,.$$

Or

$$Re\,\frac{h}{L} \qquad 1\,.$$

Since Re and h/L are both very small, it is clear that the convective terms also can be dropped. Since the pressure gradient term is retained, it is clear that

$$P \sim \frac{\mu U L}{h^2}\,.$$

The presence of the h^2 term in the denominator shows that P is quite high in this flow.

A similar analysis for the y-momentum shows that

$$\frac{\partial p}{\partial y} = 0\,.$$

The governing equations are thus Eqn. 6.1 and

$$\mu\frac{\partial^2 u}{\partial y^2} = \frac{dp}{dx}\,. \tag{7.26}$$

Consider now the flow in the slider bearing shown in Fig. 7.8. The boundary conditions for u are that $u = U$ at $y = 0$ and $u = 0$ at $y = h$. For the y-component of velocity, they are $v = 0$ at $y = 0$ or $y = h$. Also note that $p = p_\infty$ at $x = 0$ and $x = L$.

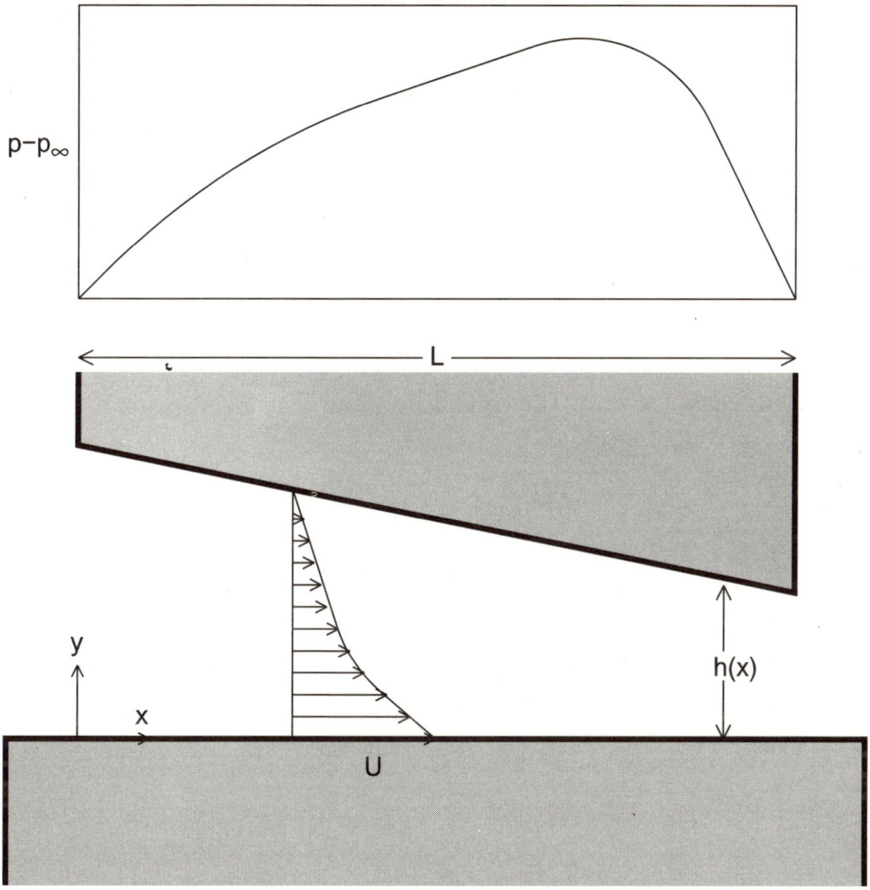

Fig. 7.8: Creeping flow in the small gap between two sliding surfaces

If we integrate Eqn. 7.26 and apply the boundary conditions, we are led to

$$u = \left(1 - \frac{y}{h}\right)\left[U - \frac{1}{2\mu}\frac{dp}{dx}yh\right] . \tag{7.27}$$

The y-component of velocity can now be determined using Eqn. 6.31. Thus

$$v = -\frac{Uy^2}{2h^2}\frac{dh}{dx} - \frac{d^2p}{dx^2}\frac{1}{2\mu}\left(\frac{y^3}{3} - \frac{y^2h}{2}\right) + \frac{dp}{dx}\frac{1}{2\mu}\frac{dh}{dx}\frac{y^2}{2}. \qquad (7.28)$$

In deriving the above equation, we have imposed $v = 0$ at $y = 0$. Therefore,

$$v|_{y=h} = -\frac{U}{2}\frac{dh}{dx} + \frac{d^2p}{dx^2}\frac{h^3}{12\mu} + \frac{dp}{dx}\frac{h^2}{4\mu}\frac{dh}{dx}$$

has not been set to zero. The volume flow rate Q at any x-station is given as

$$\begin{aligned} Q &= \int_0^{h(x)} \left(1 - \frac{y}{h}\right)\left[U - \frac{1}{2\mu}\frac{dp}{dx}yh\right] dy \\ &= \frac{1}{2}Uh - \frac{h^3}{12\mu}\frac{dp}{dx}. \end{aligned}$$

The volume rate, clearly, has to be a constant *i.e.*, it cannot be a function of x. Thus

$$\frac{dQ}{dx} = -\frac{U}{2}\frac{dh}{dx} + \frac{d^2p}{dx^2}\frac{h^3}{12\mu} + \frac{dp}{dx}\frac{h^2}{4\mu}\frac{dh}{dx} = 0\,.$$

Since the expression for dQ/dx is the same as that for $v|_{y=h}$, it is clear that $v|_{y=h} = 0$. Thus,

$$\frac{dp}{dx} = -\frac{12\mu Q}{h^3} + \frac{6\mu U}{h^2}. \qquad (7.29)$$

If we integrate both sides of this equation from $x = 0$ to $x = L$, we have

$$\int_0^L \frac{dp}{dx}\,dx = -\int_0^L \frac{12\mu Q}{h^3}\,dx + \int_0^L \frac{6\mu U}{h^2}\,dx\,.$$

Since $p = p_\infty$ at $x = 0$ and $x = L$, we get

$$Q = \frac{U}{2} \frac{\int_0^L \frac{dx}{h^2}}{\int_0^L \frac{dx}{h^3}} . \tag{7.30}$$

On the other hand, if we integrate Eqn. 7.29 from $x = 0$ to some x, we get

$$p = p_\infty + 6\mu U \int_0^x \frac{d\xi}{h^2} - 12\mu Q \int_0^x \frac{d\xi}{h^3} , \tag{7.31}$$

where ξ is a dummy integration variable. With the geometry of the gap, $h(x)$ known, the pressure distribution in the gap and the load that can be supported can be determined. We proceed to do this next for a simple geometry.

If the variation of the gap width is linear in x, then we can write

$$h = h_1 - (h_1 - h_2)\frac{x}{L} ,$$

where h_1 and h_2 are the gap widths at $x = 0$ and $x = L$ respectively. Upon substituting this profile into Eqns. 7.30 and 7.31, we get

$$Q = U \frac{h_1 h_2}{h_1 + h_2} ,$$

and

$$p = p_\infty + 6\mu U L \frac{(h_1 - h)(h - h_2)}{(h_1^2 - h_2^2)\, h^2} .$$

If $h_1 > h_2$, *i.e.*, if the gap is converging in the direction of motion, then $p > p_\infty$ inside the gap and so a load can be supported. The variation of $p - p_\infty$ along the gap is shown in Fig. 7.8 on the top. The normal force on the block(s) can be calculated as

$$\mathfrak{F} = \int_0^L pW\, dx = \frac{6\mu U W L^2}{(h_1 - h_2)^2} \left[\ln\left(\frac{h_1}{h_2}\right) - 2\frac{h_1 - h_2}{h_1 + h_2} \right] ,$$

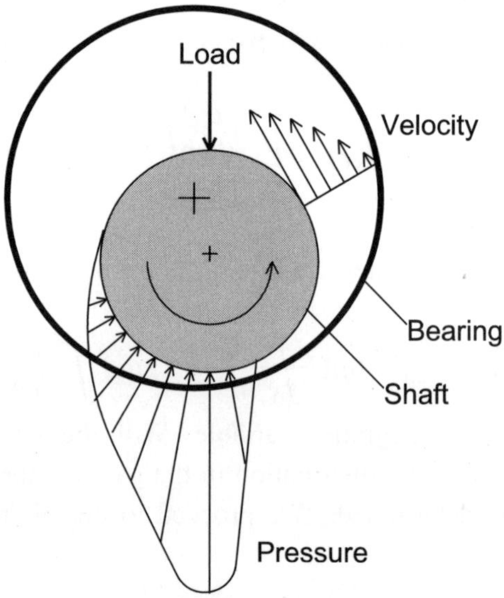

Fig. 7.9: Schematic of a journal bearing with the velocity and pressure profiles of the flow in the gap

where W is the width of the bearing in the z-direction.

The application of the theory developed above to a journal bearing is shown in Fig. 7.9. Here, the shaft is located eccentrically inside the bearing and rotates within it. The space between the bearing and the shaft is filled with a lubricating oil. There is also a downward load that acts on the shaft. Owing to the eccentric placement of the shaft, a converging gap is created and as shown above, high pressure develops in this gap. The resultant pressure force in the upward direction balances the load and the shaft is supported in the bearing without metal-to-metal contact. It is important to note that the pressure is higher only in the convergent part of the gap. The development above assumes the flow to be 2D, or, equivalently, the width of the bearing perpendicular to the page to be very large, whereas, in reality a journal bearing will have a finite width. Hence, the velocity component in the z-direction will also be non-zero. The theory discussed here can be extended to include such 3D effects.

Exercises

(1) Consider parallel flow of a fluid in the annulus between two horizontal, concentric cylinders of radius R_1 and R_2. Derive expressions for the velocity profile and the pressure drop per unit length of annulus. Compare with the corresponding expression for the flow through a pipe.

(2) Consider the flow in the channel of height $2H$ between two stationary plates. The top half of the channel is filled with a fluid of density ρ_1 and viscosity μ_1, while the lower half is filled with a fluid of density ρ_2 ($\rho_2 > \rho_1$) and viscosity μ_2. Derive an expression for the power required to maintain a volume flow rate of Q through the channel.

(3) A liquid film of height h flows down a long inclined plane at an angle α to the horizontal. Assuming the film thickness to remain constant, obtain an expression for the velocity profile inside the film. Use a coordinate system with the x and y axes oriented along and perpendicular to the inclined surface.

(4) A continuous belt passing vertically upward through a chemical bath at speed U picks up a liquid film. After some vertical distance from the surface of the bath, the thickness of the film becomes constant and equal to h. The density of the liquid is ρ and the viscosity is μ. Obtain an expression for the velocity profile inside the film in terms of the given quantities.

(5) Consider the flow of a fluid confined in the shallow cavity ($H \ll L$) as shown in the figure. The upper plate moves at a steady speed of U. Assuming that there is no leakage of fluid, derive an expression for the velocity profile along the vertical centerline of the cavity (far away from the ends). [$\frac{Uy}{H}\left(\frac{3y}{H} - 2\right)$]

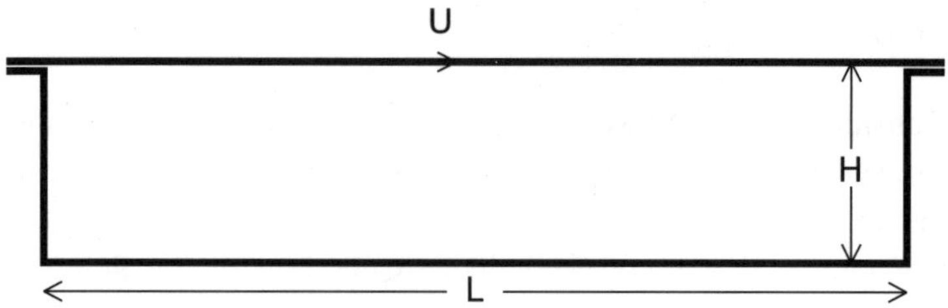

(6) Consider a minor modification to the above problem assuming that there is a constant leakage of Q, through the cavity as shown in the figure. Derive an expression for the velocity profile along the vertical centerline of the cavity (far away from the ends).

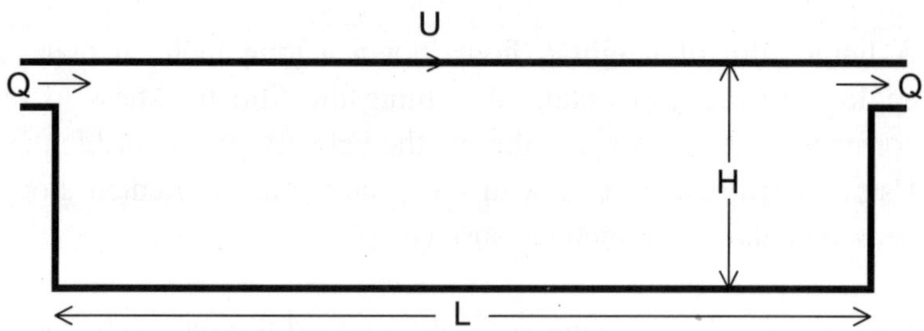

(7) Consider the arrangement shown in the figure consisting of a moving piston inside a cylinder. The cylinder is filled with an incompressible fluid of density ρ and viscosity μ. The diameter of the cylinder is D and its length is L. The piston moves to the left with a constant velocity U. The gap width h between the cylinder and the piston may be assumed to be much smaller than any of the other linear dimensions. Determine the pressure difference across the piston face and the force opposing the movement of the piston. The piston rod may be assumed to be negligibly thin.

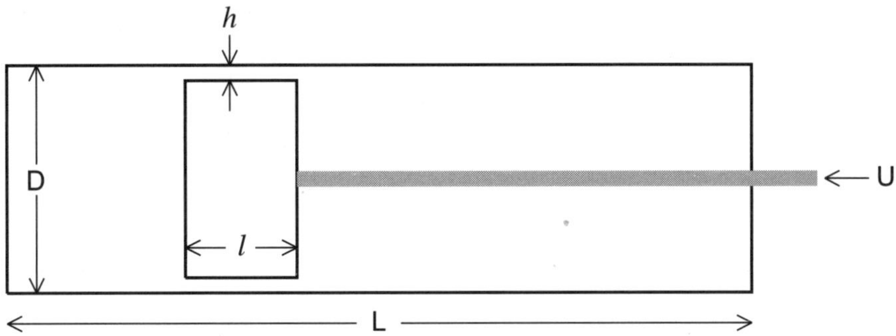

(8) An incompressible fluid of density ρ and viscosity μ is contained in the narrow space between two disks of radius R as shown in the figure. The disks are a distance h apart ($h \ll R$). The upper disk is pulled upwards very slowly with speed W. Determine the force exerted on the upper disk and the direction along which this force acts. Assume that the flow is quasi-steady and take the atmospheric pressure to be p_∞. You may assume that the z-velocity, w, is much smaller than the radial velocity, u. Gradients along the z-direction can be taken to be much larger than those in the r-direction. [$3\pi\mu W R^4/(2h^3)$ vertically downward]

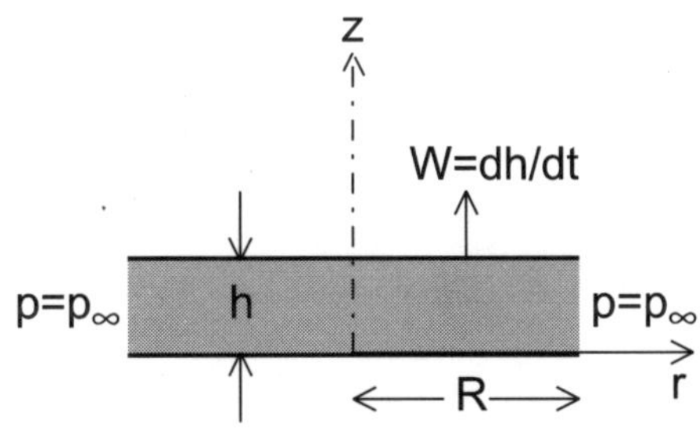

Chapter 8

Turbulent Flows

Many of the laminar flow solutions discussed in the previous chapters are not realizable in practice as they are unstable. Even the tiniest disturbance is sufficient to trigger this instability and change the flow field. As mentioned earlier, since disturbances are always present, these are the solutions that are actually seen in real life. The laminar flow solutions can be seen only under extremely controlled laboratory conditions. The Hagen-Poiseuille flow through a pipe becomes unstable at a Reynolds number around 2000 and becomes turbulent shortly afterwards, depending on the roughness of the inside surface of the pipe. Boundary layer flow over a flat plate, left to itself, transitions and becomes fully turbulent when the local Reynolds number exceeds 5×10^6. The transition Reynolds number is very sensitive to the turbulence in the freestream as well as the roughness of the surfaces. The transition process involves the appearance and amplification of two dimensional disturbances, formation of three dimensional disturbances followed by turbulent spots and, finally, coalescence of the spots into fully turbulent flow. Linear and non-linear hydrodynamic stability theories deal with most aspects of the transition process.

8.1 Reynolds averaging

Once the flow becomes turbulent, it also becomes highly unsteady. All the flow properties at any given spatial location exhibit fluctuations in their values at every instant. In addition, turbulent flows are inherently three dimensional, as alluded to above. The equations that govern turbulent flows are still Eqns. 3.11 and 3.23, since nothing was assumed about the flow being laminar or turbulent in their derivation. However, owing to the above two aspects, namely, unsteadiness and three dimensionality, closed form solutions to turbulent flows are not possible. Even numerical solutions to these equations with the so-called Direct Numerical Simulations (DNS) using the most powerful computers are extremely difficult to obtain, except in a few cases.

Reynolds (1895) proposed that, at any instant in time, a flow property be written as the sum of a mean value and a fluctuation about this mean value. By substituting this decomposition into the governing equations and then averaging them over time, equations that govern the mean motion can be derived†. The question of how to generate equations that govern the fluctuating components will be discussed subsequently.

8.2 Reynolds Averaged Navier Stokes (RANS) equations

We start by writing

$$
\begin{aligned}
u &= \bar{u} + u' \\
v &= \bar{v} + v' \\
w &= \bar{w} + w' \\
p &= \bar{p} + p' \,,
\end{aligned}
\tag{8.1}
$$

where, the mean value of a quantity f is defined as

$$\bar{f} = \frac{1}{T}\int_{t_0}^{t_0+T} f\, dt\,.$$

The averaging should be done over a sufficiently long enough period of time T so that the fluctuations are completely removed. This will ensure that $\overline{f'} = 0$‡ and hence $\bar{\bar{f}} = \bar{f}$. Other properties of this averaging process

†It must be recalled that a time varying instantaneous flow and the corresponding mean flow were discussed in detail in Section 2.2.

‡If the mean flow itself is unsteady, then the time period of averaging T must be long enough that the fluctuations about this mean are eliminated but short enough that the

are listed below:

$$\begin{aligned}
\overline{f+g} &= \bar{f}+\bar{g} \\
\overline{f\bar{g}} &= \bar{f}\,\bar{g} \\
\overline{f'\bar{g}} &= 0 \\
\overline{\frac{\partial f}{\partial \xi}} &= \frac{\partial \bar{f}}{\partial \xi} \\
\overline{fg} &= \bar{f}\bar{g}+\overline{f'g'} \\
\overline{\int f\,d\xi} &= \int \bar{f}\,d\xi\,.
\end{aligned} \tag{8.2}$$

It is important to note that while $\overline{f'} = 0$,

$$\overline{f'^2} = \frac{1}{T}\int_{t_0}^{t_0+T} f'^2\,dt \neq 0$$

since the integrand is always positive.

Substitution of the decomposition in Eqn. 8.1 into Eqns. 3.11 and 3.23 leads to the Reynolds Averaged Navier-Stokes (RANS) equations. We start with the continuity equation 3.11, which can be written as

$$\frac{\partial \bar{u}}{\partial x}+\frac{\partial u'}{\partial x}+\frac{\partial \bar{v}}{\partial y}+\frac{\partial v'}{\partial y}+\frac{\partial \bar{w}}{\partial z}+\frac{\partial w'}{\partial z}=0\,.$$

If we time average this equation and use Eqn. 8.2, we get

$$\frac{\partial \bar{u}}{\partial x}+\frac{\partial \bar{v}}{\partial y}+\frac{\partial \bar{w}}{\partial z}=0\,. \tag{8.3}$$

If we subtract this equation from the previous equation, we get

$$\frac{\partial u'}{\partial x}+\frac{\partial v'}{\partial y}+\frac{\partial w'}{\partial z}=0\,. \tag{8.4}$$

unsteadiness in the mean flow is preserved. This, of course, assumes that the fluctuations are at a higher frequency and the unsteadiness in the mean flow is at a lower frequency.

Since Eqn. 3.11 is linear, the decomposition of the velocity components in a manner given in Eqn. 8.1 yields Eqns. 8.3 and 8.4, which are identical to Eqn. 3.11 except that the instantaneous velocities are replaced by their mean and the fluctuating components respectively.

Upon substituting Eqn. 8.1 into the x-momentum equation in Eqn. 3.23, we get

$$\begin{aligned}\frac{\partial}{\partial t}(\bar{u}+u') + (\bar{u}+u')\,\frac{\partial}{\partial x}(\bar{u}+u') + (\bar{v}+v')\,\frac{\partial}{\partial y}(\bar{u}+u')\\ + (\bar{w}+w')\,\frac{\partial}{\partial z}(\bar{u}+u')\\ = -\frac{1}{\rho}\,\frac{\partial(\bar{p}+p')}{\partial x}\\ + \nu\left(\frac{\partial^2(\bar{u}+u')}{\partial x^2} + \frac{\partial^2(\bar{u}+u')}{\partial y^2} + \frac{\partial^2(\bar{u}+u')}{\partial z^2}\right).\end{aligned}$$

If we time average this equation, we get

$$\begin{aligned}\frac{\partial \bar{u}}{\partial t} + \overline{(\bar{u}+u')\,\frac{\partial}{\partial x}(\bar{u}+u')} + \overline{(\bar{v}+v')\,\frac{\partial}{\partial y}(\bar{u}+u')} + \overline{(\bar{w}+w')\,\frac{\partial}{\partial z}(\bar{u}+u')}\\ = -\frac{1}{\rho}\,\frac{\partial \bar{p}}{\partial x} + \nu\left(\frac{\partial^2 \bar{u}}{\partial x^2} + \frac{\partial^2 \bar{u}}{\partial y^2} + \frac{\partial^2 \bar{u}}{\partial z^2}\right). \qquad (8.5)\end{aligned}$$

Let us now consider the first term under the overbar. If we use the product rule in Eqn. 8.2, we get

$$\begin{aligned}\overline{(\bar{u}+u')\,\frac{\partial}{\partial x}(\bar{u}+u')} &= \overline{(\bar{u}+u')}\;\overline{\frac{\partial}{\partial x}(\bar{u}+u')} + \overline{(\bar{u}+u')'\left(\frac{\partial}{\partial x}(\bar{u}+u')\right)'}\\ &= \bar{u}\,\frac{\partial \bar{u}}{\partial x} + \overline{u'\,\frac{\partial u'}{\partial x}}\,.\end{aligned}$$

In deriving this expression, we have used the fact that $\bar{u} + u'' = u'$. The second and third terms under the overbar in Eqn. 8.5 can be written in a similar fashion. Equation 8.5 thus becomes

$$\frac{\partial \bar{u}}{\partial t} + \bar{u}\,\frac{\partial \bar{u}}{\partial x} + \bar{v}\,\frac{\partial \bar{u}}{\partial y} + \bar{w}\,\frac{\partial \bar{u}}{\partial z} + \overline{u'\,\frac{\partial u'}{\partial x}} + \overline{v'\,\frac{\partial u'}{\partial y}} + \overline{w'\,\frac{\partial u'}{\partial z}}$$
$$= -\frac{1}{\rho}\frac{\partial \bar{p}}{\partial x} + \nu\left(\frac{\partial^2 \bar{u}}{\partial x^2} + \frac{\partial^2 \bar{u}}{\partial y^2} + \frac{\partial^2 \bar{u}}{\partial z^2}\right). \quad (8.6)$$

Next, we write

$$\overline{u'\,\frac{\partial u'}{\partial x}} = \frac{\partial}{\partial x}\overline{(u'^2)} - \overline{u'\,\frac{\partial u'}{\partial x}}$$
$$\overline{v'\,\frac{\partial u'}{\partial y}} = \frac{\partial}{\partial y}\overline{(u'v')} - \overline{u'\,\frac{\partial v'}{\partial y}}$$
$$\overline{w'\,\frac{\partial u'}{\partial z}} = \frac{\partial}{\partial z}\overline{(u'w')} - \overline{u'\,\frac{\partial w'}{\partial z}}.$$

If we add these three equations, we get

$$\overline{u'\,\frac{\partial u'}{\partial x}} + \overline{v'\,\frac{\partial u'}{\partial y}} + \overline{w'\,\frac{\partial u'}{\partial z}} = \frac{\partial}{\partial x}\overline{(u'^2)} + \frac{\partial}{\partial y}\overline{(u'v')} + \frac{\partial}{\partial z}\overline{(u'w')}$$
$$- \overline{u'\,\underbrace{\left(\frac{\partial u'}{\partial x} + \frac{\partial v'}{\partial y} + \frac{\partial w'}{\partial z}\right)}_{=0}}$$
$$= \frac{\partial}{\partial x}\overline{(u'^2)} + \frac{\partial}{\partial y}\overline{(u'v')} + \frac{\partial}{\partial z}\overline{(u'w')},$$

where the term in the underbrace has been set to zero using Eqn. 8.4. Equation 8.6 can now be written as

$$\frac{\partial \bar{u}}{\partial t}+\bar{u}\,\frac{\partial \bar{u}}{\partial x}+\bar{v}\,\frac{\partial \bar{u}}{\partial y}+\bar{w}\,\frac{\partial \bar{u}}{\partial z}+\overbrace{\frac{\partial}{\partial x}(\overline{u'^2})+\frac{\partial}{\partial y}(\overline{u'v'})+\frac{\partial}{\partial z}(\overline{u'w'})}$$
$$=-\frac{1}{\rho}\frac{\partial \bar{p}}{\partial x}+\nu\left(\frac{\partial^2 \bar{u}}{\partial x^2}+\frac{\partial^2 \bar{u}}{\partial y^2}+\frac{\partial^2 \bar{u}}{\partial z^2}\right). \quad (8.7)$$

It is quite straightforward to derive similar equations from the y- and the z-momentum equations in 3.23 and these are shown below.

$$\frac{\partial \bar{v}}{\partial t}+\bar{u}\,\frac{\partial \bar{v}}{\partial x}+\bar{v}\,\frac{\partial \bar{v}}{\partial y}+\bar{w}\,\frac{\partial \bar{v}}{\partial z}+\overbrace{\frac{\partial}{\partial x}(\overline{u'v'})+\frac{\partial}{\partial y}(\overline{v'^2})+\frac{\partial}{\partial z}(\overline{v'w'})}$$
$$=-\frac{1}{\rho}\frac{\partial \bar{p}}{\partial y}+\nu\left(\frac{\partial^2 \bar{v}}{\partial x^2}+\frac{\partial^2 \bar{v}}{\partial y^2}+\frac{\partial^2 \bar{v}}{\partial z^2}\right), \quad (8.8)$$

$$\frac{\partial \bar{w}}{\partial t}+\bar{u}\,\frac{\partial \bar{w}}{\partial x}+\bar{v}\,\frac{\partial \bar{w}}{\partial y}+\bar{w}\,\frac{\partial \bar{w}}{\partial z}+\overbrace{\frac{\partial}{\partial x}(\overline{u'w'})+\frac{\partial}{\partial y}(\overline{v'w'})+\frac{\partial}{\partial z}(\overline{w'^2})}$$
$$=-\frac{1}{\rho}\frac{\partial \bar{p}}{\partial z}+\nu\left(\frac{\partial^2 \bar{w}}{\partial x^2}+\frac{\partial^2 \bar{w}}{\partial y^2}+\frac{\partial^2 \bar{w}}{\partial z^2}\right). \quad (8.9)$$

Equations 8.7-8.9 together with Eqn. 8.3 are called the Reynolds Averaged Navier-Stokes (RANS) equations. They are identical to their counterparts, Eqns. 3.11 and 3.23, except for (a) the replacement of the instantaneous velocities with the corresponding mean velocities and (b) the presence of the terms underneath the overbrace. It is clear from the above derivation that the latter is caused by the non-linear, convective terms. Had it not been for these terms, governing equations for the mean and the fluctuating quantities could have been generated in a manner similar to Eqns. 8.3 and 8.4[†]. This means that $\bar{u}$, $\bar{v}$, $\bar{w}$ and $\bar{p}$ can in principle said to be governed by Eqns. 8.3 and 8.7-8.9. However, it is not possible to generate equations that govern the six unknown quantities involving the fluctuating components that appear in Eqns. 8.7-8.9. This leaves us with an insufficient number of equations resulting in the so-called *closure problem*. It is important to understand that this difficulty is associated with the RANS equations and *not* the Navier-Stokes equations.

[†]and the study of fluid mechanics would have been far easier!

8.3 Bouissenesq hypothesis

If we rewrite Eqns. 8.7-8.9 by taking the terms in the overbraces to the right hand side and combining them with the viscous terms, we get

$$\frac{\partial \bar{u}}{\partial t} + \bar{u}\,\frac{\partial \bar{u}}{\partial x} + \bar{v}\,\frac{\partial \bar{u}}{\partial y} + \bar{w}\,\frac{\partial \bar{u}}{\partial z} = -\frac{1}{\rho}\frac{\partial \bar{p}}{\partial x} + \frac{1}{\rho}\frac{\partial}{\partial x}\left(\mu\frac{\partial \bar{u}}{\partial x} - \rho\overline{u'^2}\right) + \frac{1}{\rho}\frac{\partial}{\partial y}\left(\mu\frac{\partial \bar{u}}{\partial y} - \rho\overline{u'v'}\right) + \frac{1}{\rho}\frac{\partial}{\partial z}\left(\mu\frac{\partial \bar{u}}{\partial z} - \rho\overline{u'w'}\right), \quad (8.10)$$

$$\frac{\partial \bar{v}}{\partial t} + \bar{u}\,\frac{\partial \bar{v}}{\partial x} + \bar{v}\,\frac{\partial \bar{v}}{\partial y} + \bar{w}\,\frac{\partial \bar{v}}{\partial z} = -\frac{1}{\rho}\frac{\partial \bar{p}}{\partial y} + \frac{1}{\rho}\frac{\partial}{\partial x}\left(\mu\frac{\partial \bar{v}}{\partial x} - \rho\overline{u'v'}\right) + \frac{1}{\rho}\frac{\partial}{\partial y}\left(\mu\frac{\partial \bar{v}}{\partial y} - \rho\overline{v'^2}\right) + \frac{1}{\rho}\frac{\partial}{\partial z}\left(\mu\frac{\partial \bar{v}}{\partial z} - \rho\overline{v'w'}\right), \quad (8.11)$$

and

$$\frac{\partial \bar{w}}{\partial t} + \bar{u}\,\frac{\partial \bar{w}}{\partial x} + \bar{v}\,\frac{\partial \bar{w}}{\partial y} + \bar{w}\,\frac{\partial \bar{w}}{\partial z} = -\frac{1}{\rho}\frac{\partial \bar{p}}{\partial z} + \frac{1}{\rho}\frac{\partial}{\partial x}\left(\mu\frac{\partial \bar{w}}{\partial x} - \rho\overline{u'w'}\right) + \frac{1}{\rho}\frac{\partial}{\partial y}\left(\mu\frac{\partial \bar{w}}{\partial y} - \rho\overline{v'w'}\right) + \frac{1}{\rho}\frac{\partial}{\partial z}\left(\mu\frac{\partial \bar{w}}{\partial z} - \rho\overline{w'^2}\right). \quad (8.12)$$

There is no theoretical justification or basis for combining these terms with the viscous terms. However, doing so, allows them to be interpreted as stresses due to turbulence. Indeed, a comparison of Eqns. 8.10-8.12 with Eqns. 3.19 and 3.22 shows that the stress tensor due to turbulence can be written as

$$\bar{\bar{\Pi}}_t = \begin{pmatrix} -\rho\overline{u'^2} & -\rho\overline{u'v'} & -\rho\overline{u'w'} \\ -\rho\overline{u'v'} & -\rho\overline{v'^2} & -\rho\overline{v'w'} \\ -\rho\overline{u'w'} & -\rho\overline{v'w'} & -\rho\overline{w'^2} \end{pmatrix} . \tag{8.13}$$

This is called the Reynolds stress tensor and it is also symmetric just like $\bar{\bar{\Pi}}'$ and $\bar{\bar{\epsilon}}$. The stress strain relation, Eqn. 3.21, then becomes (after setting $\vec{\nabla} \cdot \vec{u} = 0$)

$$\bar{\bar{\Pi}}' = 2\mu\,\bar{\bar{\epsilon}} + \bar{\bar{\Pi}}_t\,.$$

If we substitute for $\bar{\bar{\epsilon}}$ from Eqn. 3.20, then

$$\bar{\bar{\Pi}}' = \begin{pmatrix} 2\mu\frac{\partial\bar{u}}{\partial x} & \mu\left(\frac{\partial\bar{v}}{\partial x} + \frac{\partial\bar{u}}{\partial y}\right) & \mu\left(\frac{\partial\bar{u}}{\partial z} + \frac{\partial\bar{w}}{\partial x}\right) \\ \mu\left(\frac{\partial\bar{v}}{\partial x} + \frac{\partial\bar{u}}{\partial y}\right) & 2\mu\frac{\partial\bar{v}}{\partial y} & \mu\left(\frac{\partial\bar{w}}{\partial y} + \frac{\partial\bar{v}}{\partial z}\right) \\ \mu\left(\frac{\partial\bar{u}}{\partial z} + \frac{\partial\bar{w}}{\partial x}\right) & \mu\left(\frac{\partial\bar{w}}{\partial y} + \frac{\partial\bar{v}}{\partial z}\right) & 2\mu\frac{\partial\bar{w}}{\partial z} \end{pmatrix}$$

$$+ \begin{pmatrix} -\rho\overline{u'^2} & -\rho\overline{u'v'} & -\rho\overline{u'w'} \\ -\rho\overline{u'v'} & -\rho\overline{v'^2} & -\rho\overline{v'w'} \\ -\rho\overline{u'w'} & -\rho\overline{v'w'} & -\rho\overline{w'^2} \end{pmatrix} .$$

A comparison of the elements of the two tensors on the right hand side of this equation immediately suggests that it will be very convenient to write the Reynolds stress tensor as

$$\bar{\bar{\Pi}}_t = \begin{pmatrix} 2\mu_t \frac{\partial \bar{u}}{\partial x} - \frac{2}{3}\rho k & \mu_t \left(\frac{\partial \bar{v}}{\partial x} + \frac{\partial \bar{u}}{\partial y} \right) & \mu_t \left(\frac{\partial \bar{u}}{\partial z} + \frac{\partial \bar{w}}{\partial x} \right) \\ \mu_t \left(\frac{\partial \bar{v}}{\partial x} + \frac{\partial \bar{u}}{\partial y} \right) & 2\mu_t \frac{\partial \bar{v}}{\partial y} - \frac{2}{3}\rho k & \mu_t \left(\frac{\partial \bar{w}}{\partial y} + \frac{\partial \bar{v}}{\partial z} \right) \\ \mu_t \left(\frac{\partial \bar{u}}{\partial z} + \frac{\partial \bar{w}}{\partial x} \right) & \mu_t \left(\frac{\partial \bar{w}}{\partial y} + \frac{\partial \bar{v}}{\partial z} \right) & 2\mu_t \frac{\partial \bar{w}}{\partial z} - \frac{2}{3}\rho k \end{pmatrix}, \quad (8.14)$$

where μ_t is a turbulent or eddy viscosity and $k = (\overline{u'^2} + \overline{v'^2} + \overline{w'^2})/2$ is the turbulence kinetic energy. The $-(2/3)\rho k$ term in the diagonal elements of the $\bar{\bar{\Pi}}_t$ is required for consistency as otherwise the trace of this tensor, $-\rho(\overline{u'^2} + \overline{v'^2} + \overline{w'^2})$ will become equal to zero by virtue of Eqn. 8.3. This *ad hoc* replacement was first suggested by Bouissenesq (1877, 1896)[†]. It resolves the closure problem, in one seemingly simple stroke, by replacing the terms involving the velocity fluctuations in the right hand side of Eqns. 8.10-8.12 with gradients of mean velocities. However, it leaves unaddressed the means by which μ_t can be calculated. This difficulty is especially acute, since μ_t is a property of the *flow* and not the fluid. Hence it varies from point to point and has to be determined along with the flow field.

8.4 Turbulence modeling

Since μ_t cannot be determined exactly, models have been developed over the years to evaluate it. The models have constants or functions which are fixed so that the predictions are consistent and agree with experimental data wherever available. Consequently, flow field predictions obtained using these models are sensitive to the values used for these constants. Although considerable efforts have been made to improve both the predictions and the reliability of these models, this shortcoming remains still. The interested reader is referred to the book *Turbulence Modeling for CFD by D. C. Wilcox, DCW Industries, La Canada, California, 1994* for an

[†]Bouissenesq, in fact, suggested that $-\rho\overline{u'v'} = \mu_t\, \partial\bar{u}/\partial y$, in the context of a boundary layer flow.

excellent exposition of various aspects of turbulence modeling.

8.5 Universal structure of the mean velocity profile in the turbulent boundary layer

In general, it is quite challenging to obtain a solution to Eqns. 8.10-8.12 even with the powerful computers available today. It is quite discouraging, if we have to resort to this strategy even for simple situations such as the turbulent flow through a pipe or over a flat plate. However, it turns out that in such situations, the quantities of interest such as pressure drop or drag force can be calculated to an acceptable degree of accuracy for engineering applications using the mean axial velocity alone. This is made especially easy by the fact that the mean axial velocity profile in a turbulent boundary layer exhibits an universal structure. This structure is discussed next. The procedure for using this in actual calculations is discussed in the next two chapters.

Klebanoff (1955), in a ground breaking experimental work, reported measurements of the mean axial velocity, the rms values of the fluctuating quantities and the Reynolds stress component $\overline{u'v'}$ in the turbulent boundary layer over a flat plate. This experimental data is shown in Fig. 8.1. The following observations can be made from the experimental data:

- To begin with, the variation of $\bar{u}/U_\infty$ near the wall is much steeper than the corresponding laminar profile (which is represented quite well by a quadratic profile). This means that the wall shear stress and hence the drag force is much higher in the turbulent boundary layer, a fact that will be established quantitatively in a later chapter.
- Although the mean flow is two dimensional, the fluctuations are three dimensional. In fact, $w'_{rms}/U_\infty = \sqrt{\overline{w'^2}}/U_\infty$, the turbulence intensity in the spanwise direction, is actually greater than that in the vertical direction.
- The turbulence intensities in all three coordinate directions, as well as the Reynolds stress component $\overline{u'v'}$ are non-zero across the entire boundary layer and become zero only outside the boundary layer.
- The turbulence intensities increase towards the wall at $y = 0$, and a

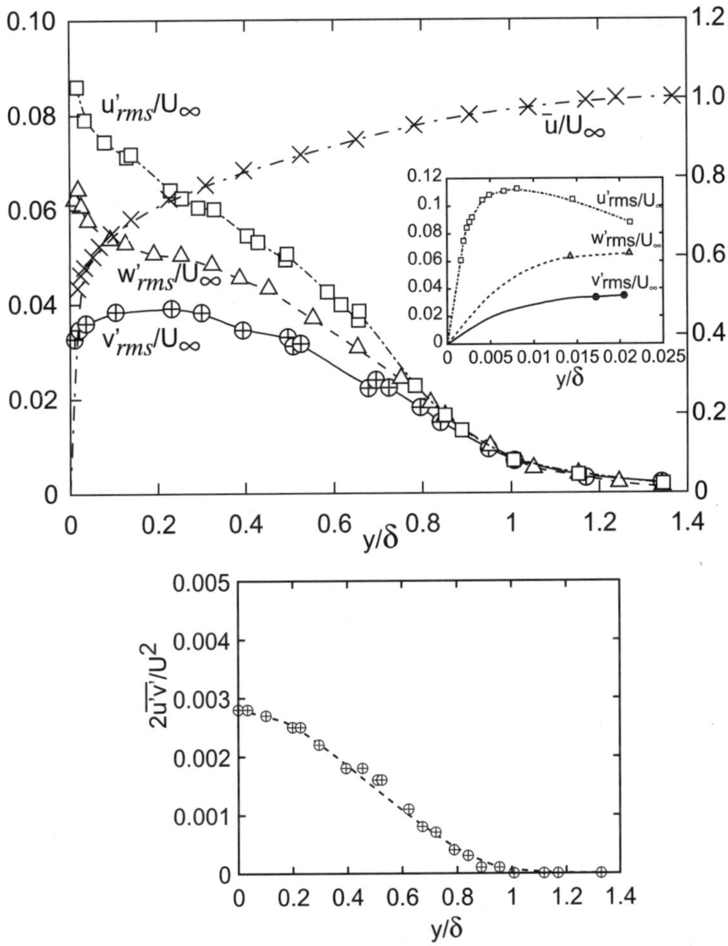

Fig. 8.1: Experimental measurements of the mean and fluctuating velocity components in the turbulent boundary layer over a flat plate. *Adapted from* P. S. Klebanoff, Characteristics of turbulence in a boundary layer with zero pressure gradient, NACA-TR-1247, 1955.

close-up view of the data given in the inset shows that they attain their peak values very close to the wall, especially u'_{rms}. Since u', v' and w' all go to zero right at the wall, it is clear that the effect due to the presence of the wall extends only for a very short distance. Similarly, the Reynolds stress component $\overline{u'v'}$ increases and reaches a peak value very close to the wall. It decreases and becomes zero at $y = 0$ over a very short distance (this is not shown in Fig. 8.1).

This seems to suggest that within the boundary layer (discussed above), there exist three regions such that,

- very close to the wall ($y/\delta < 0.005$), the viscous shear stress $\mu\partial\bar{u}/\partial y$ is very high and the turbulent shear stress $-\rho\overline{u'v'}$ is negligible,
- away from the wall ($0.005 < y/\delta < 0.01$), the viscous shear stress decreases, the turbulent shear stress increases and they are comparable in magnitude, and
- further away ($y/\delta > 0.01$), the viscous shear stress becomes negligible while the turbulent shear stress dominates.

Prandtl, von Kármán and Millikan hypothesized the existence of such regions in the 1930s in general in turbulent boundary layers. It was suggested that the turbulent boundary layer consisted of three regions (or layers):

- an inner region dominated by viscous shear stress
- an outer region dominated by turbulent shear stress
- an overlap region in which viscous and turbulent shear stress are equally important

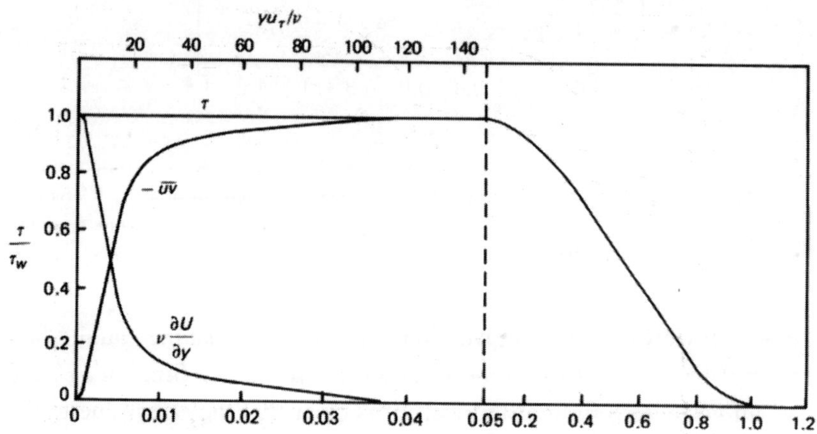

Fig. 8.2: Variation of the total shear stress, Reynolds stress and viscous stress in a turbulent boundary layer *(from Physical Fluid Dynamics by Tritton)*

The experimental data shown in Fig. 8.2 very nicely corroborates this hypothesis. The close-up view in this plot clearly shows the steep decrease in the viscous shear stress and the steep increase in the turbulent shear stress for $y/\delta < 0.01$.

The confirmation of this hypothesis, naturally results in an expectation that the mean velocity profiles in turbulent boundary layers in general should exhibit some form of similarity. However, experimental measurements of the mean flow in turbulent boundary layers shown in Fig. 8.3 for the entire range of pressure gradient fail to live up to this expectation. The only discernable similarity is that all of the profiles except the one for separating flow, show a steep variation of $\bar{u}$ near the wall.

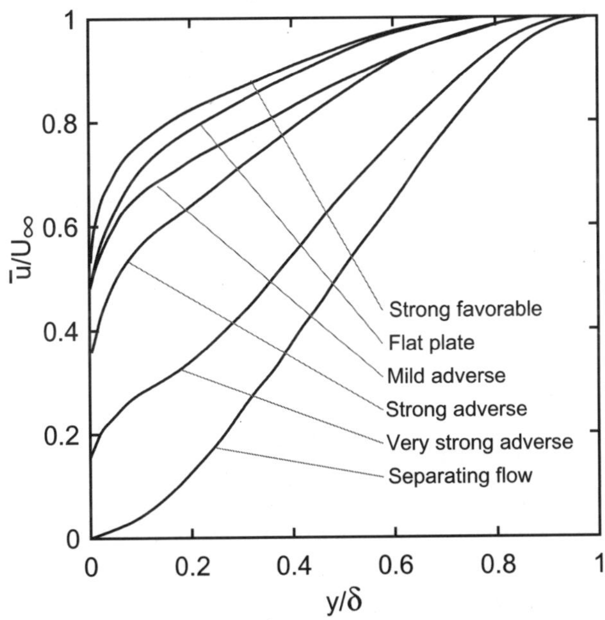

Fig. 8.3: Experimental measurements of the mean velocity in turbulent boundary layers for several flows *(adapted from Viscous Fluid Flow by White)*

It was von Kármán who suggested a replot of the data in Fig. 8.3 using different coordinates, u^+ and y^+ instead of $\bar{u}/U_\infty$ and y/δ. Here

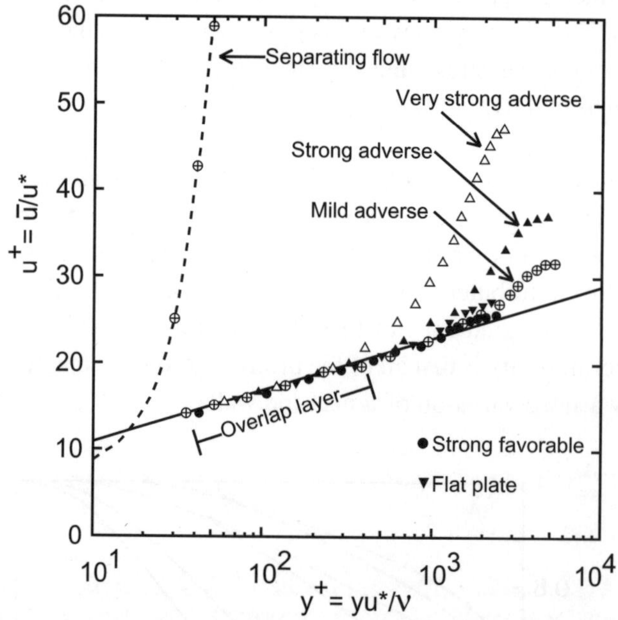

Fig. 8.4: Experimental measurements of the mean velocity in turbulent boundary layers for several flows plotted using u^+ and y^+ coordinates *(adapted from Viscous Fluid Flow by White)*

$$u^+ = \frac{\bar{u}}{u^*} \quad \text{and} \quad y^+ = \frac{yu^*}{\nu}, \tag{8.15}$$

where u^* is called the friction velocity and is defined as $u^* = \sqrt{\tau_{wall}/\rho}$.

A replot of the experimental data in Fig. 8.3 using u^+ and y^+ coordinates is shown in Fig. 8.4. Indeed, the velocity profiles for all the cases except the separating flow collapse very nicely on a single curve for $35 < y^+ < 350$. This region (or layer) is called the overlap region. The velocity profile in this region is thus a straight line in a semi-log plot. Hence

$$\frac{\bar{u}}{u^*} = \frac{1}{\kappa} \ln \frac{yu^*}{\nu} + B, \tag{8.16}$$

where κ and B are constants. κ is called the von Kármán constant. This relation is called the log-law relation. It can be seen from the straight line fit to the experimental data shown in Fig. 8.4, that $1/\kappa$ is the slope and B is the y-intercept of this line. The values for the constants are $\kappa = 0.4$ and $B = 5.5$ as suggested by Nikuradse (1930). The log-law relation can also be written in terms of $\bar{u}/U$ and y/δ as follows. If we assume the log-law region to extend all the way up to the edge of the boundary layer, then we can set $\bar{u} = U_\infty$ at $y = \delta$ in Eqn. 8.16. This gives

$$\frac{U_\infty}{u^*} = \frac{1}{\kappa} \ln \frac{\delta u^*}{\nu} + B\,,$$

Upon subtracting Eqn. 8.16 from the above expression, we get

$$\frac{U_\infty - \bar{u}}{u^*} = -\frac{1}{\kappa} \ln \frac{y}{\delta}\,.$$

In reality, the log-law relation is not applicable all the way to the edge of the boundary layer, and hence a small correction is customarily introduced into the above equation. Thus,

$$\frac{U_\infty - \bar{u}}{u^*} = -\frac{1}{\kappa} \ln \frac{y}{\delta} + A\,, \tag{8.17}$$

where the constant $A = 2.35$ for boundary layer flow and 0.65 for pipe or duct flow.

The region beyond $y^+ = 350$ is the outer (or wake) region and it can be seen the velocity profiles deviate from each other in this region. Outer laws (law of the wake) have been reasonably successful in collapsing this data by including the pressure gradient also as a parameter.

The region $0 < y^+ < 5$ is the inner layer or the viscous sub-layer. Here, the viscous stress dominates and u^+ varies linearly with y^+, *i.e.*,

$$u^+ = y^+\,. \tag{8.18}$$

The experimental data shown in Fig. 8.4 does not have any measurements

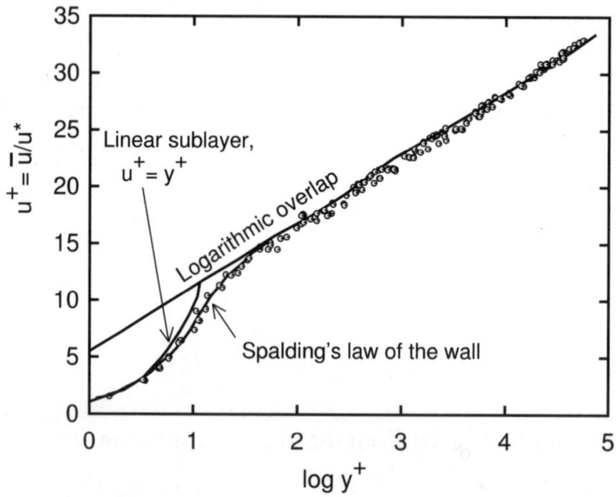

Fig. 8.5: Experimental measurements of the mean velocity in turbulent pipe flow for $4 \times 10^3 \leq Re \leq 3.2 \times 10^6$(Nikuradse, 1932 and Reichardt, 1951) *(adapted from Boundary Layer Theory by Schlichting and Gersten)*

in the viscous sub-layer. Extremely careful and painstaking experimental measurements of pipe flow carried out by Nikuradse in 1932, Reichardt in 1951 and Lindgren in 1965 validated the linear variation of the velocity in the viscous sub-layer most convincingly (Figs. 8.5 and 8.6).

The region $5 < y^+ < 35$ is called the buffer region (or layer) and the velocity u^+ transitions smoothly from the linear profile in Eqn. 8.18 to the logarithmic variation in Eqn. 8.16. Spalding (1967) derived a composite expression for u^+ that can be used all the way from the wall to the log-law region ($0 < y^+ < 350$) including the buffer layer. This is given as

$$u^+ = y^+ + e^{-\kappa B}\left[e^{\kappa u^+} - 1 - \kappa u^+ - \frac{(\kappa u^+)^2}{2} - \frac{(\kappa u^+)^3}{6}\right]. \quad (8.19)$$

This is plotted in Fig. 8.6 and it is immediately clear how well it fits the experimental data. However, since u^+ appears on both sides of the equation, it is somewhat difficult to use. Nevertheless, it is the only closed form expression that is valid up to $y^+ = 350$ and hence is usually considered as the *law of the wall*.

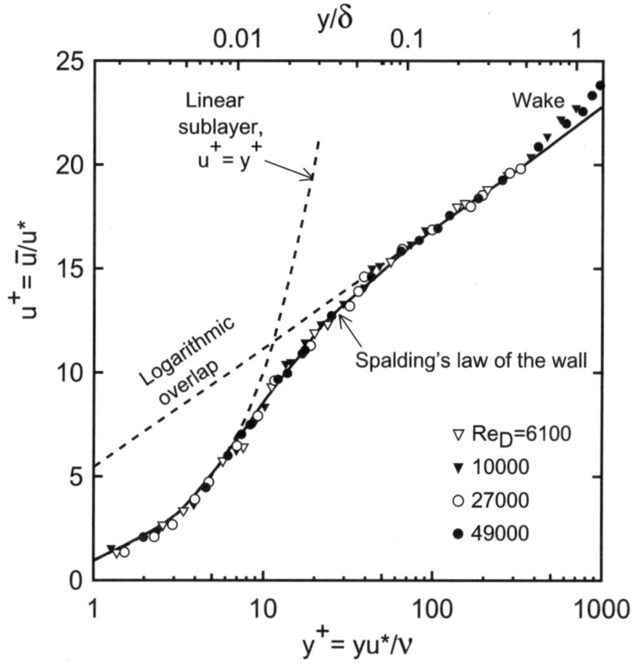

Fig. 8.6: Experimental measurements of the mean velocity in turbulent pipe flow (Lindgren, 1965) *(adapted from Viscous Fluid Flow by White)*

The experimental data plotted in Figs. 8.4 - 8.6 demonstrate clearly that the log-law region (Eqn. 8.16) and the sub-layer (Eqn. 8.18) are universal since these experiments cover both internal and external flows. Hence, they may be used with confidence for engineering calculations and this is discussed in the next two chapters.

Chapter 9

Turbulent Internal Flows

In this chapter, the universal mean velocity profile in the turbulent boundary layer is used for calculating the pressure drop, pumping power and other quantities of engineering interest in turbulent internal flows. Roughness of the conduit surface plays a major role in such flows and this is also accounted for in the development in this chapter. Some strategies for the reduction of drag in such flows are also discussed.

9.1 Turbulent flow in a pipe

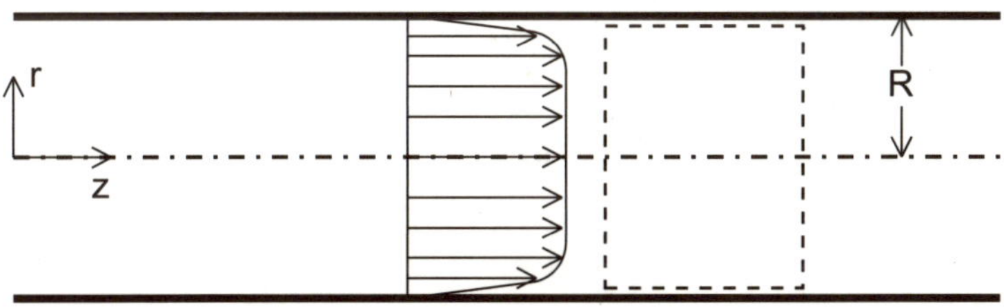

Fig. 9.1: Turbulent flow in a pipe

Consider fully developed turbulent flow through a smooth pipe as shown in Fig. 9.1. Experimental measurements for this flow shown in Figs. 8.5 and 8.6 for various Reynolds numbers reveal that there is only a negligibly small wake region (outer layer) in all the cases. This suggests that the actual velocity profile departs only slightly from the log-law profile and hence the log-law profile for the mean velocity approximates the velocity profile quite well from the wall to the centerline. Thus, the velocity profile is given as

$$\frac{w}{w^*} = \frac{1}{\kappa} \ln \frac{(R-r)w^*}{\nu} + B\,, \tag{9.1}$$

where Eqn. 8.16 has been modified for the coordinate system shown in Fig. 9.1. As before, we take $\kappa = 0.4$ and $B = 5.5$. The overbar on w which denotes that this is the mean and not the instantaneous velocity has been dropped as there is no danger of ambiguity. The velocity profile is sketched qualitatively in Fig. 9.1. It can be seen by comparing this with the laminar profile (Fig. 7.5) that the gradient near the wall and hence the wall shear stress are much higher in the case of turbulent flow. Furthermore, since the velocity profile resembles a plug flow profile, the ratio of the average to the maximum velocity can be expected to be higher than the 0.5 (and closer to 1) for the laminar case.

The maximum velocity occurs at the centerline and is given as

$$w_{max} = w^* \left(2.5 \ln \frac{Rw^*}{\nu} + 5.5\right)\,. \tag{9.2}$$

The volumetric flow rate is

$$\begin{aligned} Q &= \int_0^R w \times 2\pi r\, dr \\ &= \int_0^R w^* \left[\frac{1}{\kappa} \ln \frac{(R-r)w^*}{\nu} + B\right] 2\pi r\, dr \\ &= \pi R^2 w^* \left(\frac{1}{\kappa} \ln \frac{Rw^*}{\nu} + B - \frac{3}{2\kappa}\right) \\ &= \pi R^2 w^* \left(2.5 \ln \frac{Rw^*}{\nu} + 1.75\right)\,. \end{aligned}$$

The average velocity

$$\bar{w} = \frac{Q}{\pi R^2} = w^* \left(2.5 \ln \frac{Rw^*}{\nu} + 1.75\right)\,. \tag{9.3}$$

An integral analysis using the control volume shown in Fig. 9.1 will show that, in the case of turbulent flow also, the pressure gradient Λ that drives the flow should exactly match the pressure drop due to friction at the wall. Moreover, Eqns. 7.4 and 7.18 are applicable in the case of turbulent flow also. The latter equation is especially useful, since it allows w^* $(= \sqrt{\tau_{wall}/\rho})$ to be eliminated from Eqn. 9.3. Thus,

$$\frac{\bar{w}}{w^*} = \sqrt{\frac{\rho \bar{w}^2}{\tau_{wall}}} = \sqrt{\frac{8}{f}}. \tag{9.4}$$

Similarly

$$\frac{Rw^*}{\nu} = \frac{1}{2}\frac{D\bar{w}}{\nu}\frac{w^*}{\bar{w}} = \frac{1}{2} Re_D \sqrt{\frac{f}{8}}.$$

Upon substituting these two expressions into Eqn. 9.3, we get

$$\sqrt{\frac{8}{f}} = 2.5 \ln\left(\frac{1}{2} Re_D \sqrt{\frac{f}{8}}\right) + 1.75 \tag{9.5}$$

$$\frac{1}{f^{1/2}} = 0.884 \ln \frac{1}{2\sqrt{8}} + 0.884 \ln Re_D \sqrt{f} + 0.6187$$

$$= 0.884 \ln\left(Re_D \sqrt{f}\right) - 0.9132$$

$$= 2.0355 \log\left(Re_D \sqrt{f}\right) - 0.9132 .$$

This is the expression for the friction factor in a turbulent flow. This was first derived by Prandtl and he adjusted the constants so that it read

$$\frac{1}{f^{1/2}} = 2 \log\left(Re_D \sqrt{f}\right) - 0.8 . \tag{9.6}$$

This expression is Prandtl's universal law of friction for turbulent flow in smooth pipes. Since it is derived using the universal log-law relation, there is no restriction on the upper value of the Reynolds number and it is valid for arbitrarily large Reynolds numbers. Experimental measurements by Nikuradse in 1932 showed that the agreement between the values for f from experiments and Eqn. 9.6 is excellent up to $Re_D = 3.4 \times 10^6$. Friction factors calculated using Eqn. 9.6 at several Reynolds numbers are given in Table 9.1.

Table 9.1: Friction factor for turbulent pipe flow at different Reynolds numbers

Re_D	4000	10^4	10^5	10^6	10^7	10^8
f	0.039916	0.0308891	0.017993	0.011647	0.008103	0.005941
$\bar{w}/w_{max}$	0.7906	0.8110	0.8490	0.8748	0.8934	0.9073

The friction factor decreases with Re_D. However, as Eqns. 7.16 and 7.18 show, the head loss due to friction and the wall shear stress increase with Re_D since $\bar{w}^2$ varies as Re_D^2 for a given pipe diameter and viscosity.

From Eqns. 9.2 and 9.3, it follows that

$$\begin{aligned}
\frac{\bar{w}}{w_{max}} &= \frac{2.5\,\ln\left(Rw^*/\nu\right) + 1.75}{2.5\,\ln\left(Rw^*/\nu\right) + 5.5} \\
&= \frac{2.5\ln\left(\frac{1}{2}\,Re_D\sqrt{\frac{f}{8}}\right) + 1.75}{2.5\ln\left(\frac{1}{2}\,Re_D\sqrt{\frac{f}{8}}\right) + 5.5} \\
&= \frac{\sqrt{8/f}}{\sqrt{8/f} + 3.75}\,,
\end{aligned}$$

where, we have used Eqn. 9.5 for the final simplification. Values of $\bar{w}/w_{max}$ at different Reynolds numbers are given in Table 9.1. These values are greater than the 0.5 for the laminar case (as surmised earlier), and also demonstrate that the velocity profile becomes fuller (*i.e.*, resembles a plug flow profile) as the Reynolds number increases.

The velocity profile as written in Eqn. 9.1 is not in a convenient form owing to the presence of the friction velocity, w^*. The detailed experimental investigation by Nikuradse in 1932 showed that the velocity profile can be expressed using a power law as

$$\frac{w}{w_{max}} = \left(\frac{R-r}{R}\right)^{1/n} . \tag{9.7}$$

The value of n is obtained from a curve fit of the experimental data. The values of n as calculated by Nikuradse are given in Table 9.2. Although this form is more convenient, the dependence of the constant n on the Reynolds number, in contrast to the profile in Eqn. 9.1, which is universal, makes it less aesthetic.

Table 9.2: Values of the power law exponent n for different Reynolds numbers

Re_D	4000	2.3×10^4	1.1×10^5	1.1×10^6	2.0×10^6	3.2×10^6
n	6.0	6.6	7.0	8.8	10	10

Similarly, Eqn. 9.6 is not in a convenient form to use since f occurs on both sides of the equation. Blasius, in the year 1911 provided a curve fit to the experimental data available till then in a convenient form:

$$f = 0.3164 Re_D^{-1/4}, \qquad 4000 \leq Re_D \leq 10^5 . \tag{9.8}$$

Using this expression, the head loss can be calculated from Eqn. 7.16 as

$$h_f = \frac{fL\bar{w}^2}{2gD} = \frac{8fLQ^2}{\pi^2 gD^5} = \frac{0.0261LQ^2}{Re_D^{1/4} D^5}$$
$$= \frac{0.0277LQ^{1.75}\mu^{0.25}}{\rho^{0.25} D^{4.75}}.$$

The pressure drop is then given as

$$\Delta p = \rho g h_f = \frac{0.272LQ^{1.75}\rho^{0.75}\mu^{0.25}}{D^{4.75}}$$
$$= \frac{0.272LQ^{7/4}\rho^{3/4}\mu^{1/4}}{D^{19/4}}. \tag{9.9}$$

The corresponding expression for laminar flow from Eqn. 7.14 is given as

$$\Delta p = \frac{128\mu QL}{\pi D^4} = \frac{40.74\mu QL}{D^4} = \frac{40.74LQ^{4/4}\mu^{4/4}}{D^{16/4}}.$$

A comparison of these two expressions suggests the following (for $4000 \leq Re_D \leq 10^5$):

- The pressure drop is independent of the density of the fluid in laminar flows, while it increases as $\rho^{3/4}$ in turbulent flows.
- The dependence of the pressure drop on the viscosity of the fluid is weaker in turbulent flows
- The dependence of the pressure drop on the flow rate is stronger in the case of turbulent flows. The variation is almost Q^2 in a turbulent flow compared to Q in a laminar flow.
- The dependence of the pressure drop on the diameter is stronger in turbulent flows. Doubling the diameter reduces the pressure drop by a factor of 27 in a turbulent flow compared to 16 in the case of a laminar flow.

Example 9.1. Repeat Example 7.1 assuming the pipe to be smooth.

Solution. The Reynolds number based on the pipe diameter was calculated to be 56128. From Eqn. 9.6 the friction factor can be calculated as

$f = 0.020366$. The friction factor is higher now by a factor of nearly 18 when compared to the value obtained under the laminar assumption. With $\bar{w} = 1.19\ m/s$, $D = 1.22\ m$, $L = 1280\ kms$ and $\rho = 930\ kg/m^3$, we get from Eqn. 7.16,

$$\Delta p = \rho g h_f = \frac{\rho f L \bar{w}^2}{2D} = 14.1 \times 10^6\ N/m^2.$$

The pumping power is $\mathfrak{P} = \Delta p\, Q = 19.6\ MW$. □

In practical applications involving transport of liquids (or gases) over long distances, multiple pumping stations are used along the length of the pipe. Although the overall power required remains the same, this strategy allows the pumping to be accomplished with multiple smaller pumps rather than a single huge pump. The pressure variation along the pipe is thus saw tooth shaped rather than a single straight line (of slope $\Delta p/L$). The slope of the inclined portion of the saw tooth profile, however, is equal to $\Delta p/L$.

From published information, the maximum pressure in the Trans Alaska pipe line is about 80 $bars$. The pressure in the line must always be above atmospheric pressure (say 2 $bars$) to ensure that the gases dissolved in the crude oil stay dissolved and do not start boiling away. The distance between pumping stations, l, that is required to achieve this can be calculated as

$$\frac{(80-2)10^5}{l} = \frac{\Delta p}{L} = \frac{14.1 \times 10^6}{1280 \times 10^3} \quad \Rightarrow \quad l = 708\ kms$$

In reality, the distance between pumping stations is only about 180 kms, owing to the elevation changes along the route.

9.2 Effect of roughness

Roughness of the pipe surface has a tremendous effect on the friction factor when the flow is turbulent, in contrast to the laminar flow situation, where roughness plays no role at all. Nikuradse, one of Prandtl's students, performed a series of very systematic, extensive and careful experiments† in 1933 to investigate the effect of roughness. He glued sand grains of fixed sizes on the inside surface of a smooth pipe to simulate the effect of wall roughness‡ and measured the pressure drop as well as the velocity profiles. In this case, it is reasonable to expect the friction factor to depend on the roughness (characterized by the grain size ϵ), in addition to the Reynolds number, Re_D. Nikuradse's work showed that this dependance can be classified into three regimes:

- Hydraulically smooth regime. In this regime, the roughness is such that $\epsilon^+ = \epsilon w^*/\nu \leq 5$. The roughness is so small that it is contained within the viscous sub-layer. Therefore, roughness does not play a role at all and thus the friction factor depends only on the Reynolds number $f = f(Re_D)$ as given in Eqn. 9.6.
- Transitional regime. In this regime, $5 < \epsilon^+ \leq 70$. The wall roughness is big enough to protrude partly above the viscous sub-layer. The overall pressure drop in the pipe is a combination of viscous drag as in a smooth pipe and form drag due to the protuberances. The friction factor thus depends on the roughness as well as the Reynolds number, *i.e.*, $f = f(\epsilon/D, Re_D)$.
- Fully rough regime. In this regime, $\epsilon^+ > 70$. The roughness is large enough to protrude entirely above the sub-layer, which, as a result is completely broken up. The pressure drop is thus due to the form drag of the roughness alone. Hence the friction factor, $f = f(\epsilon/D)$.

Nikuradse's measurements also revealed that the velocity profile in rough pipes is a simple modification of the log-law profile, Eqn. 9.1, for a smooth pipe:

$$\frac{w}{w^*} = \frac{1}{\kappa} \ln \frac{(R-r)w^*}{\nu} + B + \Delta B\,, \tag{9.10}$$

where ΔB for each of the three regimes mentioned above has to be

†To quote Schlichting

‡Strictly speaking, this simulates the so-called sand grain roughness. Actual roughness on the pipe surface will usually not bear an exact resemblance. This distinction is unimportant for our purposes.

determined from experimental data. Nikuradse's experimental data shows clearly that

$$\Delta B = \begin{cases} 0, & \epsilon^+ \leq 5 \ \text{(smooth)} \\ 3.5 - 2.5 \ln \frac{\epsilon w^*}{\nu}, & \epsilon^+ > 70 \ \text{(fully rough)}. \end{cases} \tag{9.11}$$

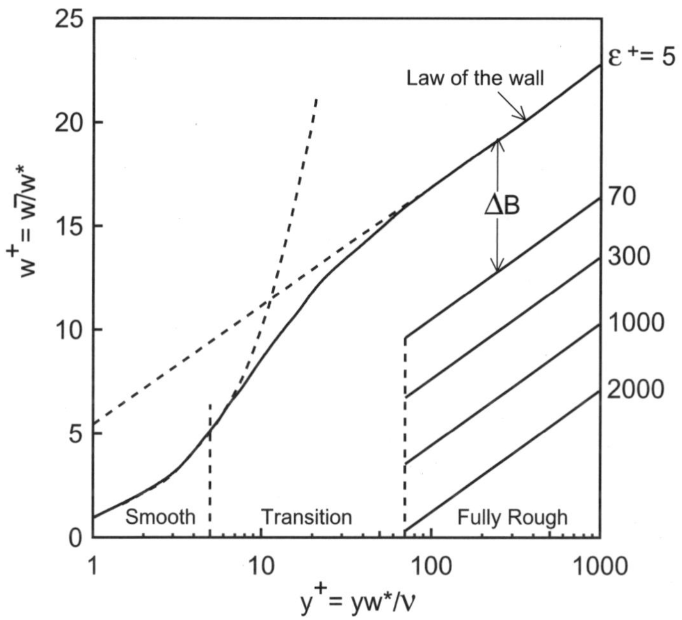

Fig. 9.2: Universal velocity profile in turbulent flow through smooth and rough pipes

Thus, in the fully rough regime, the effect of roughness is to cause a vertical downshift of the log-law profile as illustrated in Fig. 9.2. Furthermore, upon substituting for ΔB from Eqn. 9.11 into Eqn. 9.10, we get

$$\frac{w}{w^*} = \frac{1}{\kappa} \ln \frac{(R-r)w^*}{\nu} + B + 3.5 - 2.5 \ln \frac{\epsilon w^*}{\nu}$$

$$= 2.5 \ln \frac{R-r}{\epsilon} + 8.5 \tag{9.12}$$

in the fully rough regime. The kinematic viscosity ν has completely disappeared demonstrating that the drag is independent of the Reynolds number and dependent only on the roughness.

The volumetric flow rate in this case can be obtained as

$$Q = \pi R^2 w^* \left(2.5 \ln \frac{R}{\epsilon} + 4.75\right) .$$

Therefore the average velocity $\bar{w} = Q/\pi R^2$, can be written as

$$\frac{\bar{w}}{w^*} = 2.5 \ln \frac{R}{\epsilon} + 4.75 .$$

Upon combining this with Eqn. 9.4, we are led to

$$\begin{aligned}
\sqrt{\frac{8}{f}} &= 2.5 \ln \frac{R}{\epsilon} + 4.75 \\
\frac{1}{f^{1/2}} &= 0.884 \ln \frac{R}{\epsilon} + 1.68 \\
&= 2.035 \log \frac{R}{\epsilon} + 1.68 .
\end{aligned}$$

A better fit with Nikuradse's experimental data is obtained if the constant 1.68 were replaced with 1.74. Also, it is customary to replace the coefficient 2.035 of the first term with 2. Thus

$$\begin{aligned}
\frac{1}{f^{1/2}} &= 2 \log \frac{R}{\epsilon} + 1.74 \\
&= 2 \log \frac{3.7D}{\epsilon} \\
&= -2 \log \frac{\epsilon/D}{3.7} ,
\end{aligned} \tag{9.13}$$

where we have written $1.74 = 2 \log 7.4$. This is the expression for the friction factor in the fully rough regime.

9.3 Moody's chart

Equations 9.6 and 9.13 allow us to calculate the friction factors in the smooth and the fully rough regimes. There is no equation as such for the calculation of the friction factor in the transitional regime. Colebrook (1939) proposed an expression that covers all three regimes by combining Eqns. 9.6 and 9.13:

$$\frac{1}{f^{1/2}} = -2 \log \left(\frac{\epsilon/D}{3.7} + \frac{2.51}{Re_D f^{1/2}} \right), \tag{9.14}$$

where we have written $0.8 = 2 \log 2.51$. Moody (1944) calculated friction factors using this expression as well as Eqn. 7.17 and plotted them in what has since been called the Moody's chart. Equation 9.14 can also be used for calculating friction factors in non-circular geometries by using the hydraulic diameter.

Friction factors calculated from Eqn. 9.14 and 7.17 are plotted in Fig. 9.3 for laminar and turbulent pipe flows and for smooth and rough pipes. For a given ϵ/D, it can be seen that f depends initially on Re_D in the transitional regime and then becomes independent of Re_D in the fully rough regime. The dashed line in Fig. 9.3 demarcates these two regimes.

Example 9.2. Apartments in a 5 storey residential building are supplied with water from a 20000 liter overhead tank located on the roof of the building. The distance between the floors is 4.5 m and the tank is at a height of 4.5 m from the roof. The incoming pipe to the tank has a diameter of 2 $inches$. Determine the pumping power required, if the tank is to be filled in 1 hour assuming the pipe to be made of (a) PVC with $\epsilon = 0.0015\ mm$ and (b) GI with $\epsilon = 0.15\ mm$. Take $\nu = 10^{-6}\ m^2/s$ for water.

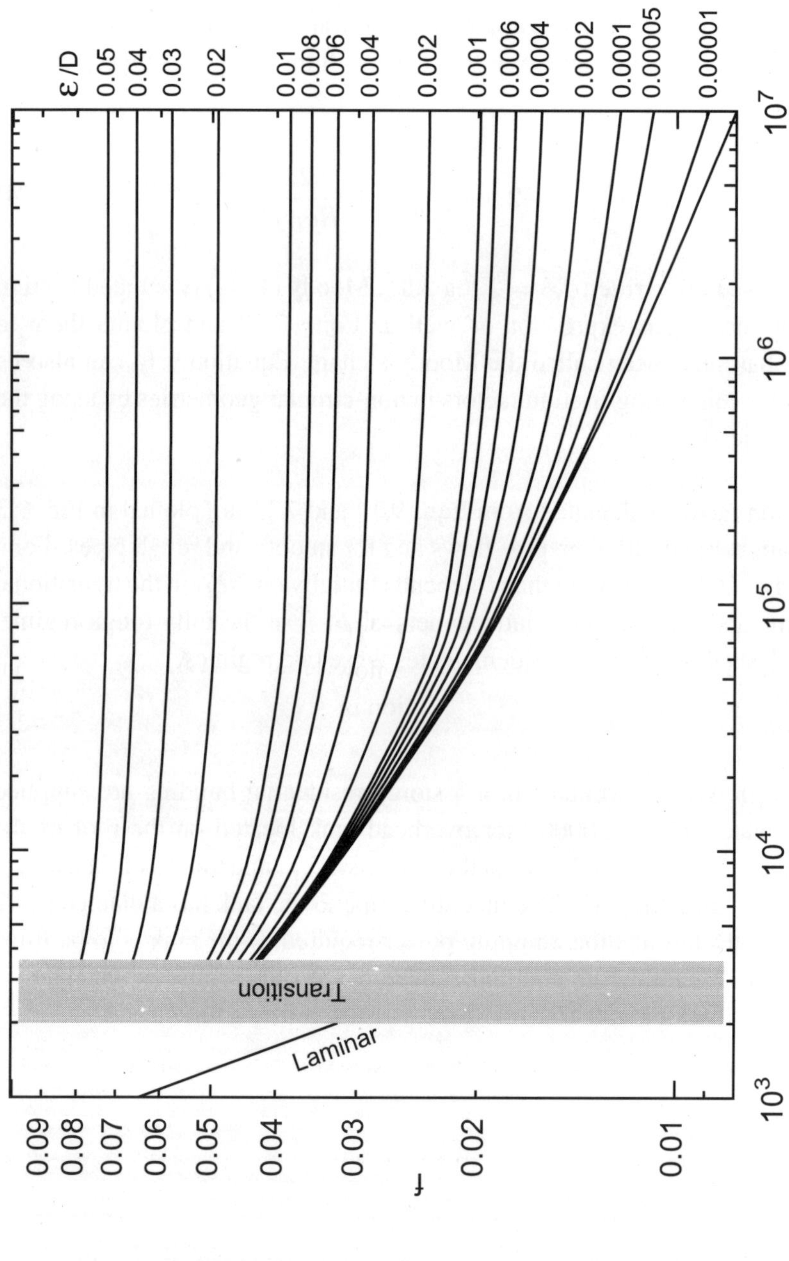

Fig. 9.3: Moody's chart for pipe flow

Solution. (a) The required flow rate is $Q = 20000 \times 10^{-3}/(3600) = 0.00556\, m^3/s$.
The average velocity is $\bar{w} = Q/(\pi D^2/4) = 0.00556 \times 4/(\pi(2 \times 0.0254)^2) = 2.743\, m/s$.
The Reynolds number is $Re_D = \bar{w}D/\nu = 2.743 \times 2 \times 0.0254/10^{-6} = 13934$.
For the PVC pipe, $\epsilon/D = 0.0015 \times 10^{-3}/(2 \times 0.0254) = 3 \times 10^{-5}$.
From Moody's chart, we see that the flow is the transitional regime and $f = 0.028$. With the length of the pipe $L = 5 \times 4.5 + 4.5 = 27\, m$, the head loss h_f is given as

$$h_f = \frac{fL\bar{w}^2}{2gD} = \frac{0.028 \times 27 \times 2.743^2}{2 \times 9.81 \times 2 \times 0.0254} = 5.7\, m\,.$$

The total head required is $h = h_f + 5 \times 4.5 + 4.5 = 32.7\, m$. The pumping power can be calculated using Eqn. 7.20 as

$$\mathfrak{P} = \rho g h Q = 1000 \times 9.81 \times 32.7 \times 0.00556 = 1783.6\, W = 2.4\, HP\,.$$

(b) For the GI pipe, $\epsilon/D = 0.15 \times 10^{-3}/(2 \times 0.0254) = 3 \times 10^{-3}$.
From Moody's chart, we see that the flow is the transitional regime and $f = 0.033$. The head loss due to friction is

$$h_f = \frac{fL\bar{w}^2}{2gD} = \frac{0.033 \times 27 \times 2.743^2}{2 \times 9.81 \times 2 \times 0.0254} = 6.73\, m\,.$$

The total head required is $h = h_f + 5 \times 4.5 + 4.5 = 33.73\, m$. The pumping power can be calculated as

$$\mathfrak{P} = \rho g h Q = 1000 \times 9.81 \times 33.73 \times 0.00556 = 1839.8\, W = 2.5\, HP\,.$$

Although ϵ/D is different by two orders of magnitude between the two pipes, the pumping power is almost identical since the flow rate is quite small. □

9.4 Drag reduction

Since the transport of liquids and gases through pipelines over long distances is a widely encountered engineering application, reduction of the drag and hence the pumping power is of tremendous practical interest. We will mention here a few techniques that have been successfully used. The readers are encouraged to search the literature for further details of these or other techniques.

In the transport of heavy oils, heating the oil prior to pumping can be helpful as it lowers the viscosity of the oil. Of course, the dependence of the pressure drop on viscosity is weak in turbulent flows as Eqn. 9.9 shows. Moreover, in the fully rough regime, the pressure drop is *independent* of the viscosity as Eqn. 9.13 shows. The energy required for heating the oil has to be debited against the saving in pumping power. Hence, the viability of this strategy depends on the actual reduction in viscosity that can be achieved by heating.

Another strategy for reducing the drag is to introduce an annular layer of a lighter, lower viscosity, immiscible fluid between the fluid that is being pumped (which forms the core) and the internal surface of the pipe. Such a flow is called a core-annular flow. For instance, water, whose viscosity is nearly two orders of magnitude less can be used as the annular fluid while pumping crude oil. The interface between the two liquids is prone to instability and hence additional chemical agents have to be added to suppress the interfacial instability.

Yet another strategy that has been shown to reduce the drag is the introduction of particles made of long chain polymers into the flow. It is speculated that the action of the turbulent shear stresses on these particles breaks them apart, resulting in a reduction in the wall shear stress and the drag.

Exercises

(1) Evaluate $\bar{w}/w_{max}$ for the Reynolds numbers in Table 9.2 using the power law profile Eqn. 9.7 and compare with the value obtained using the profile given in Eqn. 9.1.

(2) Develop an expression for the friction factor for turbulent flow between two smooth, parallel plates. Assume that the velocity profile is given by the log-law Eqn. 8.16. Compute the friction factors for $Re = 10^4,\ 10^5,\ 10^6,\ 10^7$ and 10^8 and compare them with those obtained from Eqn. 9.6 using the hydraulic diameter.

(3) Consider Couette flow between two parallel plates as shown in Fig. 7.1. Assuming the flow to be turbulent and that log-law, Eqn. 8.16, is valid, determine the dimensionless wall shear stress, $\tau_w/(\mu U/H)$, on the walls for $Re = UH/\nu = 10^6$. Compare this with the corresponding expression for laminar flow. [$\tau_{w,turb}/\tau_{w,lam} = 139$] Hint: The velocity profile is S-shaped with $u = U/2$ at $y = 0$.

(4) Chennai city receives about 190 mld (million liters per day) from Veeranam lake located 230 kms away. The water is pumped through a pipeline of 1.8 m diameter, made of steel reinforced with cement mortar. Determine the pumping power required assuming (a) the pipe to be smooth and (b) rough with ϵ=3 mm. [0.11 MW, 2.62 MW]

(5) Straws are usually used for straightening the flow entering the test section in wind tunnels. In a particular application, each straw is 30 cm in length and 5 mm in diameter and is of negligible thickness. The wind tunnel is circular in cross-section with a diameter of 100 cm and air ($\rho = 1.2\ kg/m^3$, $\nu = 1.5 \times 10^{-5}\ m^2/s$) enters with a velocity of 30 m/s. Determine the pressure drop across the straws. [972 Pa]

Chapter 10

Turbulent External Flows

Turbulent external flows are encountered in many practical applications. For instance, the flow around automobiles, trains, aircraft, ships and submarines are all turbulent external flows, just to name a few. In such applications, it is important to be able to calculate the forces that act on the body, in particular, the lift and the drag force. In this chapter, we will focus our attention primarily on the calculation of the drag force due to the friction at the wall *i.e.,* wall shear stress due to the turbulent boundary layer. In contrast, our attention was focussed on calculating the pressure drop due to friction at the wall in the case of internal flows. With this objective in mind, in this chapter, turbulent flow over a flat plate with zero pressure gradient is discussed first. Separation of the boundary layer due to a non-zero (and adverse) pressure gradient is discussed next. In this case, pressure drag is also present in addition to the friction drag.

10.1 Turbulent boundary layer over a flat plate with zero pressure gradient

Consider flow over a flat plate as sketched in Fig. 6.5, but now assuming the flow to be fully turbulent from the leading edge onwards. We wish to obtain the local skin friction coefficient, c_f, and the coefficient of drag, C_d for different Reynolds numbers. In addition, it is desirable to know the variation of the boundary layer thickness with Reynolds number. Several strategies are available to achieve these objectives. The strategies differ in the degree of difficulty and the range of applicability of the results. We will discuss a simple one in detail and confine ourselves to quoting the results for the more difficult ones.

The simplest strategy is to start from Eqn. 6.23 using an appropriate (guessed) velocity profile.

$$\mu \left.\frac{\partial u}{\partial y}\right|_{\substack{x = x_B \\ y = 0}} = \rho U_\infty^2 \frac{d}{dx} \int_0^{\delta(x_B)} \frac{u_{x=x_B}}{U_\infty} \left(1 - \frac{u_{x=x_B}}{U_\infty}\right) dy \,. \quad (6.23)$$

This equation is applicable in this case also, since nothing was assumed about the nature of the flow while deriving it. The guessed velocity profile should allow both sides of this equation to be calculated and hopefully, allow c_f to be determined. The most suitable guess for the velocity profile is the polynomial profile given in Eqn. 9.7, which, for this case can be written as†,

$$\frac{u}{U_\infty} = \left(\frac{y}{\delta}\right)^{1/n} \,. \quad (10.1)$$

It is customary to take $n = 7$. This restricts the range of applicability of the results to moderately high Reynolds numbers, or, $Re_\delta < 10^5$, from Table 9.2. The integral on the right hand side of Eqn. 6.23 is nothing but the momentum thickness. With $n = 7$, this can be evaluated as $\delta_m = (7/72)\delta$‡. However, there is a difficulty in calculating the left hand side of Eqn. 6.23 owing to the fractional exponent in Eqn. 10.1. We circumvent this difficulty by combining Eqn. 9.8 (which is valid only for $Re_\delta < 10^5$) and Eqn. 7.18 to get an expression for the wall shear stress as follows:

$$\frac{\tau_{wall}}{\rho U_\infty^2} = 0.03325 Re_\delta^{-1/4} \,. \quad (10.2)$$

†The radius of the pipe, R is replaced with the boundary layer thickness, δ and the mean velocity $\bar{w}$ is replaced with the freestream velocity, U_∞.

‡For the general case,

$$\delta_m = \frac{n}{(n+1)(n+2)} \delta \,.$$

Upon substituting these expressions into Eqn. 6.23, we get

$$\frac{7}{72}\frac{d\delta}{dx} = 0.03325\left(\frac{\nu}{U_\infty \delta}\right)^{1/4} .$$

If we integrate this equation, we get

$$\delta = 0.4275\,\frac{\nu^{1/5}x^{4/5}}{U_\infty^{1/5}} , \tag{10.3}$$

where, we have taken δ to be zero at $x = 0$. If we recall that the laminar boundary layer grows as $x^{1/2}$, it is clear that the turbulent boundary layer grows much more rapidly. Substituting Eqn. 10.3 into the right hand side of Eqn. 10.2 and rearranging gives,

$$c_f = \frac{\tau_{wall}}{\frac{1}{2}\rho U_\infty^2} = \frac{0.0822}{Re_x^{1/5}} . \tag{10.4}$$

The above expression, it may be recalled, is valid for $Re_\delta < 10^5$, or, alternatively, $Re_x < 5 \times 10^6$, if we substitute for δ from Eqn. 10.3. Since the boundary layer over a flat plate with zero pressure gradient transitions at $Re_x = 5 \times 10^5$, the range of validity of Eqn. 10.4 can be taken as $5 \times 10^5 < Re_x < 5 \times 10^6$. A comparison of Eqn. 10.4 with Eqn. 6.17 shows that the wall shear stress at a given location in the case of turbulent flow is mich higher, as expected.

The coefficient of drag for a plate of length L and width W can be obtained from Eqn. 6.22 as

$$\begin{aligned} C_d &= \frac{\mathfrak{D}}{\frac{1}{2}\rho U_\infty^2 WL} = \frac{7}{36}\frac{\delta}{L} \\ &= \frac{0.083125}{Re_L^{1/5}} , \qquad 5 \times 10^5 < Re_L < 5 \times 10^6 . \end{aligned} \tag{10.5}$$

As stated earlier, we have assumed the flow to be turbulent right from the leading edge in this development. If the flow is laminar from the leading edge up to the transition point and becomes turbulent afterwards, then Eqns. 10.4 and 10.5 have to be modified. In this case, the drag force on the plate is given as

$$\mathfrak{D} = \begin{pmatrix} \text{Turbulent} \\ \text{drag} \\ \text{from} \\ x = 0 \text{ to} \\ x = L \end{pmatrix} - \begin{pmatrix} \text{Turbulent} \\ \text{drag} \\ \text{from} \\ x = 0 \text{ to} \\ x = x_{trans} \end{pmatrix} + \begin{pmatrix} \text{Laminar} \\ \text{drag} \\ \text{from} \\ x = 0 \text{ to} \\ x = x_{trans} \end{pmatrix},$$

where $x = 0$ is the leading edge of the plate and x_{trans} is the location at which transition occurs. Since the laminar drag is less than the turbulent drag, there is a reduction in the net drag force in this case. Upon substituting for the turbulent and laminar drag from Eqns. 10.5 and 6.20, we get

$$\begin{aligned} \mathfrak{D} = {} & C_{d,t}\left(\frac{1}{2}\rho U_\infty^2\right) W\,L \\ & - C_{d,t}\left(\frac{1}{2}\rho U_\infty^2\right) W\,x_{trans} \\ & + C_{d,l}\left(\frac{1}{2}\rho U_\infty^2\right) W\,x_{trans}\,, \end{aligned}$$

where the subscripts l and t denote laminar and turbulent respectively. The coefficient of drag for this case can thus be written as

$$\begin{aligned} C_d &= \frac{\mathfrak{D}}{\frac{1}{2}\rho U_\infty^2 W L} \\ &= C_{d,t} - \frac{x_{trans}}{L}\left(C_{d,t} - C_{d,l}\right) \\ &= C_d, t - \frac{Re_{trans}}{Re_L}\left(C_{d,t} - C_{d,l}\right) \\ &= C_{d,t} - \frac{A}{Re_L}\,, \end{aligned} \tag{10.6}$$

where

$$A = Re_{trans}\left(C_{d,t} - C_{d,l}\right) \tag{10.7}$$

and Re_{trans} is the transition Reynolds number. In reality, transition occurs in the range $3 \times 10^5 < Re_{trans} < 3 \times 10^6$.

In many applications, Re_L can be as high as 10^9 and hence the limitation on the Reynolds number in conjunction with Eqns. 10.4 and 10.5 is rather severe. Since the log-law relation, Eqn. 8.16, is valid for arbitrarily large Reynolds numbers, the aforementioned limitation can be eliminated, if we use it in place of Eqn. 10.1. The experimental data for flat plate shown in Fig. 8.4 exhibits little or no wake at all, and so the log-law relation can be used all the way from the wall to the edge of the boundary layer. If we let $u = U_\infty$ at $y = \delta$ in Eqn. 8.16, we are led to

$$\frac{U_\infty}{u^*} = \frac{1}{\kappa} \ln \frac{\delta u^*}{\nu} + B\,. \tag{10.8}$$

Since $u^* = \sqrt{\tau_{wall}/\rho}$, the following relations also hold:

$$\frac{U_\infty}{u^*} = \sqrt{\frac{U_\infty^2 \rho}{\tau_{wall}}} = \sqrt{\frac{2}{c_f}}\,, \tag{10.9}$$

and

$$\frac{\delta u^*}{\nu} = \frac{\delta}{\nu} \sqrt{\frac{\tau_{wall}}{\rho}} = Re_\delta \sqrt{\frac{c_f}{2}}\,.$$

If we substitute these two relations into Eqn. 10.8, we are immediately led to

$$\sqrt{\frac{2}{c_f}} = \frac{1}{\kappa} \ln\left(Re_\delta \sqrt{\frac{c_f}{2}} \right) + B\,. \tag{10.10}$$

Schlichting recommends $\kappa = 0.3936$ and $B = 5.56$ for the flat plate. Equation 10.10 can be used in principle to determine the c_f without any restriction on the Reynolds number. However, it cannot be used as it stands, since the variation of δ with x is not known. This can be determined by following the same procedure that we adopted earlier, that is, by substituting the velocity profile, Eqn. 8.16, into the integral momentum equation, Eqn. 6.23. The algebra is tedious, but, at the end, both c_f and C_d can be calculated for any Re_L. Schlichting provided curve fits to the data as follows:

$$c_f = \frac{1}{(2\log Re_x - 0.65)^{2.3}}, \tag{10.11}$$

$$C_d = \frac{0.455}{(\log Re_L)^{2.58}}. \tag{10.12}$$

If the boundary layer is laminar initially, the coefficient of drag can be modified in the same manner as before. Thus,

$$C_d = \frac{0.455}{(\log Re_L)^{2.58}} - \frac{A}{Re_L}. \tag{10.13}$$

This is called the Prandtl-Schlichting formula. The quantity A can be evaluated using Eqn. 10.7. Values of C_d calculated using this formula agree with experimental data to within a few percent.

Both Eqns. 10.4 and 10.11 show that the wall shear stress decreases along the plate, due to the increasing thickness of the boundary layer. Equations 10.5 and 10.12 show that C_d also decreases along the length of the plate. The drag force on the plate, however, increases with the length of the plate owing to the increase in the wetted area.

As mentioned in Chapter 8, the log-law is applicable only in the overlap layer and not in the viscous sub-layer (or the outer layer *i.e.,* the wake region). As Fig. 8.4 clearly shows, the log-law is not applicable all the way to the wall. On the other hand, Spalding's equation, Eqn. 8.19 is applicable all the way from the wall to the outer edge of the overlap layer, as shown in Fig. 8.6. In fact, White† showed that it is possible to obtain better agreement with experimental data by using Eqn. 8.19, instead of the log-law expression. The outcome of such an analysis was the following

†see Viscous Fluid Flow by F. M. White

explicit relation:

$$c_f = \frac{0.455}{[\ln 0.06 Re_x]^2}\,. \tag{10.14}$$

This formula gives excellent predictions for Re_x up to 10^{10}.

10.1.1 *Effect of roughness*

The effect of roughness on the local skin friction coefficient, c_f and the coefficient of drag, C_d can be accounted for in the same manner as that of internal flows. In the latter case, the roughness was characterized by ϵ/D or, equivalently, ϵ/R. Three regimes, demarcated by the value of $\epsilon^+ = \epsilon w^*/\nu$, were also identified. These observations can be carried over to the present case as well, with minor modifications. For instance, the pipe radius, R is replaced by the boundary layer thickness δ and $\epsilon^+ = \epsilon u^*/\nu$. However, it is important to realize that R is a constant whereas, δ increases with x. Consequently, the effect of the roughness changes character along the length of the plate, although ϵ itself remains constant. From Eqn. 10.9, we get $u^* = U_\infty \sqrt{c_f/2}$. Since c_f decreases with x, u^* and hence ϵ^+ also decrease with x. Thus, the flow may be in the fully rough regime near the leading edge of the flat plate (assuming the flow to be turbulent from the leading edge), transition and then become hydraulically smooth towards the trailing edge.

Schlichting and Prandtl showed that, in this case also, Nikuradse's sand roughness data could be used with a shift that depended on ϵ^+. They suggested the following expressions for calculating c_f and C_d in the fully rough regime:

$$c_f = \frac{1}{[2.87 + 1.58\log(x/\epsilon)]^{2.5}}\,, \qquad 10^2 < \frac{x}{\epsilon} < 10^6 \tag{10.15}$$

$$C_d = \frac{1}{[1.89 + 1.62\log(L/\epsilon)]^{2.5}}\,, \qquad 10^2 < \frac{L}{\epsilon} < 10^6\,. \tag{10.16}$$

As before in the case of pipe flow, here too, the dependence on Reynolds number disappears in the fully rough regime.

Example 10.1. A supertanker (Fig. 10.1) that transports crude oil has a length of 400 m, a width of 100 m and a draft of 25 m. Determine the power required to overcome friction drag and cruise at a speed of 12 $knots$, assuming the wetted surface of the ship to be (a) smooth and (b) rough with $\epsilon = 10\ mm$. For seawater, $\nu = 1.4 \times 10^{-6}\ m^2/s$ and $\rho = 1020\ kg/m^3$. Assume $Re_{trans} = 5 \times 10^5$.

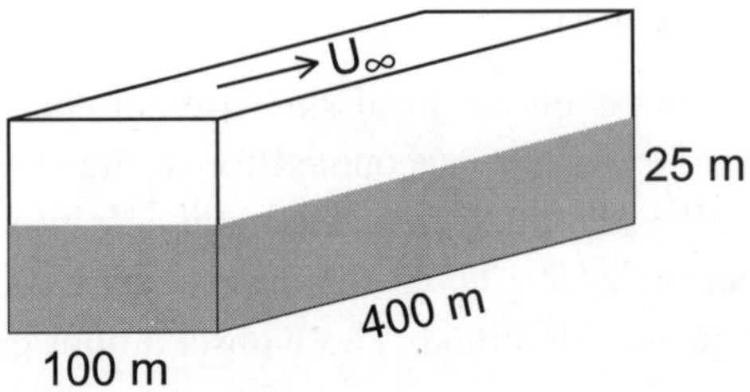

Fig. 10.1: Illustration for Example 10.1. Wetted area is shown in gray.

Solution. The total wetted area is $400(100 + 2 \times 25)m^2$, if we ignore the contribution from the forward and rear sides which are normal to the freestream. Hence, the supertanker can be modelled as a flat plate of length $400m$ and width $150m$.

(a) With the given values, we can get

$$\begin{aligned} Re_L &= \frac{U_\infty L}{\nu} \\ &= 12\ knots \times 0.51444\ \frac{m/s}{knots} \times 400\ m \times \frac{1}{1.4 \times 10^{-6}}\ \frac{s}{m^2} \\ &= 1.76 \times 10^9 \end{aligned}$$

From Eqn. 10.7,

$$\begin{aligned} A &= 5 \times 10^5 \left(\frac{0.455}{(\log Re_{trans})^{2.58}} - \frac{1.328}{Re_{trans}^{1/2}} \right) \\ &= 5 \times 10^5 (5.11 \times 10^{-3} - 1.88 \times 10^{-3}) = 1616 \end{aligned}$$

where we have used Eqns. 10.12 and 6.20. It follows from Eqn. 10.13 that $C_d = 1.46 \times 10^{-3}$. Therefore the drag force

$$\begin{aligned}\mathfrak{D} &= C_d \times \frac{1}{2}\rho U_\infty^2 LW \\ &= 1.46 \times 10^{-3} \times \frac{1}{2} \times 1020 \times (12 \times 0.5144)^2 \times (400) \times (150)\, N \\ &= 1.7 \times 10^6\, N\end{aligned}$$

The power required to overcome friction drag is $\mathfrak{P} = \mathfrak{D} U_\infty = 10.5\, MW \approx 14000\, hp$.

(b) We can assume the flow to be fully rough along the entire length since $L/\epsilon = 40000 < 10^6$. Therefore, from Eqn. 10.16, $C_d = 3.75 \times 10^{-3}$. The drag force and the power are $3.75/1.46 = 2.57$ times the values obtained in part (a) □

10.2 Turbulent flows with non-zero pressure gradient

The derivation of closed form expressions for c_f and C_d in the previous section was possible due to the fact that the pressure gradient was zero. In addition, since the flat plate was aligned with the freestream direction, the drag force acting on it was entirely due to skin friction. Unfortunately, the pressure gradient is non-zero in the majority of external flows of practical interest. The net drag force in such flows is no longer just skin friction drag, but a combination of pressure drag and skin friction drag†.

Similar to its laminar flow counterpart, the turbulent boundary layer also separates in the presence of an adverse pressure gradient. However, the turbulent boundary layer is able to withstand an adverse pressure gradient much better, owing to the momentum infused by the fluctuations into the instantaneous streamwise velocity component. As seen in Fig. 8.1, the magnitudes of the fluctuations relative to the freestream velocity (the so-called turbulence intensity) are quite high even very close to the wall. It may be recalled that the fluctuations diminish only in the viscous sub-layer. Consequently, separation is delayed in the case of turbulent boundary

†This is in contrast to internal flows such as flow through pipes and ducts, where the drag force is exclusively due to skin friction.

layers. This is illustrated in Fig. 10.2 for the flow around a circular cylinder.

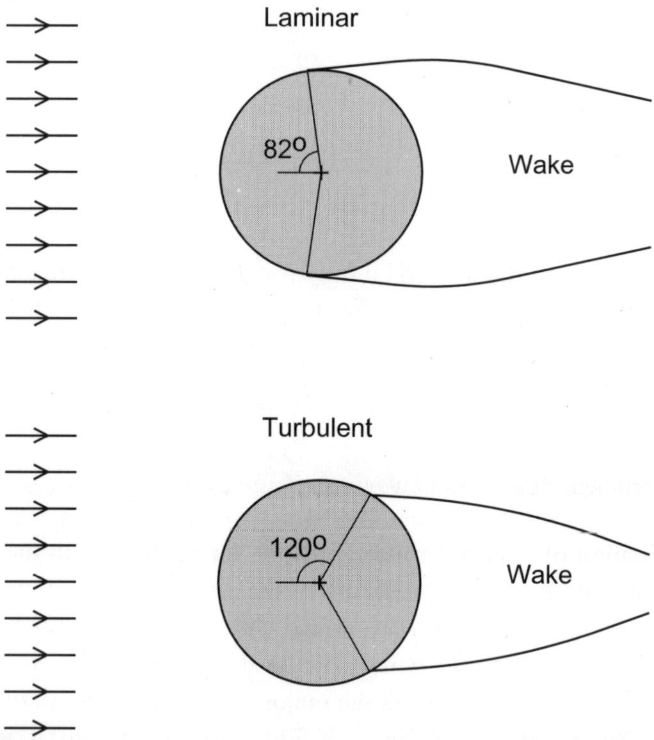

Fig. 10.2: Separation of the boundary layer for laminar and turbulent flow around a cylinder

The separation point moves from $\theta = \pm 82°$ to $\theta = \pm 120°$ when the boundary layer transitions from laminar to turbulent. The wake, which was quite broad becomes small and compact†. This has important consequences for the flow around bluff bodies, where the pressure drag dominates. Since the pressure drag in such cases arises from the difference in the pressure between the front and the rear of the body, a reduction in

†For objects such as the square cylinder in Fig. 2.4, the separation locations are fixed *i.e.,* the flow always separates from the corners and so there is no change in the size of the wake due to transition.

the size of the wake as seen in Fig. 10.2 can cause a dramatic reduction in the pressure drag. The friction drag, however, is higher when the boundary layer is turbulent, since the velocity gradient and hence the wall shear stress are both high. Notwithstanding the increase in the friction drag, the total drag decreases due to boundary layer transition owing to the large reduction in the pressure drag. This is usually referred to as the *drag crisis*. This phenomenon is seen only in external flows around bluff bodies without fixed separation points. Golf balls, for instance, have dimples to exploit this reduction in drag and increase the travel distance. As the Reynolds number increases beyond the transition value, the net drag increases as the pressure drag remains more or less constant and the friction drag increases.

It should be clear from the above discussion that, it is not possible to obtain closed form expressions for C_d for most of the external flows, regardless of whether the flow is laminar or turbulent. Experiments or numerical simulations are required to evaluate C_d (and hence the drag force, which is the quantity of primary interest in most engineering applications) and these values are available in Fluid Mechanics handbooks. It is customary to define C_d of an object due to external flow around it as

$$C_d = \frac{\mathfrak{D}}{\frac{1}{2}\,\rho U^2 \mathcal{A}}, \tag{10.17}$$

where $\mathfrak{D}$ is the drag force, ρ is the density of the fluid, U is the magnitude of the relative velocity between the object and the fluid and $\mathcal{A}$ is the area of the object projected normal to the freestream flow direction. The drag force acts in such a manner as to diminish the magnitude of the relative velocity between the object and the fluid. The following examples illustrate the use of C_d in external flows.

Example 10.2. Consider a filter composed of an array of cylindrical fibers (Fig. 10.3). The filter is 0.05 m thick and each fiber has a diameter of 0.2 mm. There are N fibers per unit cross-section. For air flowing through the filter at 2 m/s, determine the pressure drop for $N = 1000/cm^2$. For air, take $\rho = 1.2255\ kg/m^3$ and $\mu = 1.78 \times 10^{-5}\ kg/(m.s)$. C_d for each fiber may be taken to be 2.8.

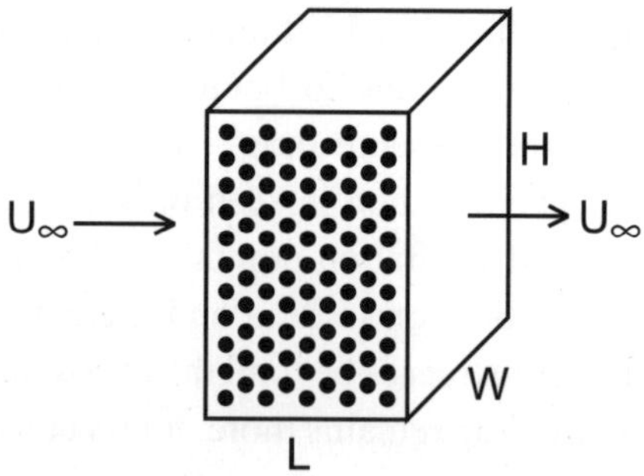

Fig. 10.3: Illustration for Example 10.2

Solution. With reference to Fig. 10.3, a momentum balance for the control volume surrounding the filter can be written as

$$0 = (\Delta p)\, H \times W - N \times L \times H\, (\mathfrak{D})\,.$$

Here, $\mathfrak{D}$ is the drag force on each fiber. The left hand side of this equation is zero as there is no change in the momentum of the fluid across the filter. From the definition of C_d, we can write

$$\mathfrak{D} = C_d \times \frac{1}{2}\rho U_\infty^2 W \times d\,,$$

where d is the diameter of the fiber. Thus,

$$\Delta p = N \times L C_d \times \frac{1}{2}\rho U_\infty^2 \times d\,.$$

If we substitute the given values, we get

$$\begin{aligned}\Delta p &= 1000 \times 10^4\, \frac{cm^2}{m^2} \times 0.05 m \times 2.8 \\ &\times \frac{1}{2} \times 1.2255 \frac{kg}{m^3} \times 2^2\, \frac{m^2}{s^2} \times 0.2 \times 10^{-3}\, m \\ &= 686.28 \frac{N}{m^2} \square\end{aligned}$$

Example 10.3. Determine the terminal velocity of a spherical raindrop of 100 μm diameter as it falls vertically through still air. For air, take

$\rho = 1.2255\,kg/m^3$ and $\mu = 1.78 \times 10^{-5}\,kg/(m.s)$. For water, take $\rho = 1000\,kg/m^3$. The following correlation for C_d due to Chow (An Introduction to Computational Fluid Dynamics, Wiley, 1980) may be used:

$$C_d = \begin{cases} 24/Re_D & Re_D \leq 1 \\ 24/Re_D^{0.646} & 1 < Re_D \leq 400 \\ 0.5 & 400 < Re_D \leq 3 \times 10^5 \\ 0.000366 Re_D^{0.4275} & 3 \times 10^5 < Re_D \leq 2 \times 10^6 \\ 0.18 & Re_D > 2 \times 10^6 \end{cases}$$

Solution. The forces that act on the raindrop as it falls, are its weight (acting downwards), buoyancy (acting upwards) and drag (acting upwards). Since the density of water is three orders of magnitude higher than that of air, the buoyancy force can be neglected. The equation of motion of the raindrop can thus be written as

$$m\frac{d^2x}{dt^2} = mg - \mathfrak{D} = mg - C_d \times \frac{1}{2}\rho_{air}U^2 \times \frac{\pi D^2}{4},$$

where m is the mass of the raindrop, D is its diameter, x is its displacement in the downward direction and $U = dx/dt$ is its velocity. Initially, the velocity of the raindrop and hence the drag force is small. The raindrop accelerates primarily under the influence of the downward force due to its weight. As the velocity of the raindrop increases, the drag force acting on it also increases. Eventually the two terms in the right hand side of the above equation balance each other and the acceleration of the raindrop then becomes zero. Thereafter the raindrop falls with a constant velocity called the terminal velocity. This can be calculated by setting d^2x/dt^2 equal to zero in the above equation. Thus

$$C_d \times \frac{1}{2}\rho_{air}U^2 \times \frac{\pi D^2}{4} = mg = \rho_{water} \times \frac{4}{3}\frac{\pi D^3}{8}g.$$

Or

$$U = \sqrt{\frac{4}{3}\frac{\rho_{water}}{\rho_{air}}\frac{gD}{C_d}}.$$

This has to be solved iteratively since C_d also depends on U through Re_D. We start with a guessed value for U. This allows us to calculate Re_D and hence C_d. U can then be calculated from the above equation. The calculation can be stopped once the guessed and the calculated value for U agree.

$U\ (m/s)$ (guessed)	1	10	0.5	0.3	0.2	0.25
$U\ (m/s)$ (calculated)	0.4	0.83	0.31	0.27	0.23	0.25
Re_D	6.88	68.8	3.44	2.06	1.38	1.72
C_d	6.9	1.5593	10.8	15.02	19.52	16.9

In this example, the flow around the raindrop turns out to be laminar as indicated by the low value for Re_D □

10.3 Drag reduction in external flows

Reduction of drag in turbulent external flows is extremely important as it can result in considerable savings in fuel consumption of automobiles, trains, aircraft and ocean going vessels. A few strategies for reducing the pressure drag have already been discussed in section 6.4. Some techniques with proven reduction of skin friction drag in turbulent flows are†:

- Large eddy breakup devices (LEBU)
- Riblets
- Suction and blowing

Large eddy breakup devices are thin plates or ribbons oriented lengthwise along the flow direction and suspended near the outer edge of the turbulent layer. Large vortical structures that are formed here usually are broken

†See the excellent article *The fix for tough spots* by Anders, Walsh, and Bushnell that appeared in *Aerospace America* in January, 1988

down into smaller and smaller sizes by the action of turbulence and they exert considerable influence over the wall shear stress. The LEBU devices breakup these vortices as they form, thereby lowering the skin friction. Under laboratory conditions, these devices have been shown to reduce the drag by as much as 10% in the skin friction. In actual applications, these ribbons have to be attached to the surface and can be envisaged to form a "ring" around the surface. The drag associated with the devices themselves can negate some of the gains achieved in the overall drag reduction.

Fig. 10.4: A close-up image of longitudinal riblets. *Courtesy: NASA*

Riblets[†] are corrugations (or grooves or striations) which are aligned lengthwise to the flow direction as shown in Fig. 10.4. The dimensions of the riblets (peak-to-valley) are of the order of 0.006 inches for subsonic freestream speeds and about 0.001 inches for supersonic speeds. These

[†]There is a very interesting biological origin to the research on riblets. Sharks possess very fine scales on their skin which have been shown to result in lesser drag than a smooth skin. See the excellent review article *Fluid Mechanics of Biological Surfaces and their Technological Application* by Bechert, Bruse, Hage and Meyer that appeared in Naturwissenschaften, Vol. 87, 2000, pp. 157-171.

dimensions roughly translate to y^+ of 10-15. The spanwise spacing (peak-to-peak) is usually of the same order *i.e.,* y^+ of 10-15. It is thus clear that the riblets operate within or just beyond the viscous sub-layer, in contrast to the LEBUs. In fact, to the naked eye these would appear to be just scratches on the surface. Anders *et al.,* speculate that "the reduction in drag is due to the enhancement of transverse (spanwise) viscous forces by strong, spanwise surface geometry gradients. These forces, in turn, produce a relatively quiescent flow in the riblet valley that pushes skin-friction-producing turbulence up and away from the surface". Although the wall shear stress is reduced, the riblets do increase the wetted area almost by a factor of 2. An inspection of the viscous drag term in Eqn. 6.38 shows that the drag force can increase due to the increased wetted surface area albeit the reduction in τ_{wall}. Hence, the geometry of the riblets must be chosen with care.

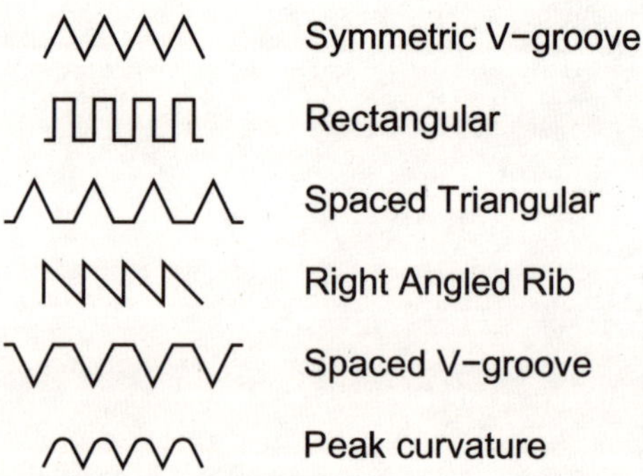

Fig. 10.5: Cross-sections of different riblet geometries. *Adapted from* M. J. Walsh, Riblets for aircraft skin friction reduction, NASA-N88-14955.

A cross-sectional view of some possible riblet geometries are shown in Fig. 10.5. The symmetric V-groove geometry has been found to be the most optimum with a drag reduction of 8% or so, over that of a smooth surface. Although the grooves can be machined on the surface, use of an adhesive backed vinyl sheet etched with the riblet geometry, has been found to be more practical. It is of interest to note that the riblets can be used in internal flows also to achieve drag reduction. Anders *et al.,* also report that, since

LEBUs and riblets operate in different parts of the boundary layer, they can be used in conjunction to achieve an enhanced reduction in the drag.

Suction (removal of fluid near the surface) and blowing (injection of high momentum fluid) have already been discussed in Section 6.4. Removing the fluid near the surface reduces the boundary layer thickness and causes a relaminarization of the boundary layer. Since the skin friction drag is higher in a turbulent boundary layer than a laminar one, this results in a reduction in the drag. Injection of a high momentum fluid energizes the boundary layer and also results in a lower skin friction.

Exercises

(1) The range of validity of Eqn. 10.5 can be greatly improved by using Eqn. 10.10 and the values for n given in Table 9.2 for the higher Reynolds numbers. Using this procedure, determine C_d for the highest three Reynolds numbers in Table 9.2 and compare them with the values obtained from Eqn. 10.13.

(2) Determine the terminal velocity of a steel sphere 10 mm in diameter ($\rho = 7850\ kg/m^3$) when falling through (a) water ($\rho = 1000\ kg/m^3,\ \ \nu = 10^{-6}\ m^2/s$) and (b) oil ($\rho = 870\ kg/m^3,\ \ \nu = 1.195 \times 10^{-4}\ m^2/s$). [8.46 m/s, 4.52 m/s]

(3) A torpedo is shaped in the form of a cylinder 55 cm in diameter with a hemispherical front cap and a total length of 5 m. Determine the power required to overcome skin friction drag when the torpedo travels in sea water ($\rho = 1025\ kg/m^3,\ \ \nu = 1.05 \times 10^{-6}\ m^2/s$) at a speed of 45 knots (23 m/s). Assume that the flow is attached along the entire length of the surface and take ϵ= 0.5 mm. [0.2 MW]
By taking the C_d based on frontal area to be 0.45, determine the power required to overcome pressure drag. [0.67 MW]

(4) A box shaped van with a streamlined front has the following dimensions: 2.5 m wide, 4 m tall and 10 m long. Determine the power required to overcome friction drag when the van travels at a speed of 100 km/hr. Assume that the boundary layer is turbulent and attached throughout. [3.3 MW]
By taking the C_d based on frontal area to be 0.45, determine the power required to overcome pressure drag. [57.87 MW]

(5) A cylindrical chimney is 2 m in diameter and 60 m high. What is the bending moment at the foot of the chimney when it is exposed to winds at 90 km/hr. The C_d based on frontal area may be taken as 0.95. [1.2825 MN.m]

Suggested Reading

Elementary Fluid Dynamics by D. J. Acheson, Oxford University Press, 1996.

Fundamental Mechanics of Fluids by I. G. Currie, 3rd Edition, Marcel Dekker, 2005.

Boundary Layer Theory by H. Schlichting and K. Gersten, Eighth Edition, Springer 2000.

Physical Fluid Dynamics by D. J. Tritton, Oxford University Press, 1988.

Fluid Mechanics by F. M. White, Fifth Edition, McGraw-Hill Higher Education, 2003.

Viscous Fluid Flow by F. M. White, Second Edition, McGraw-Hill International Editions, 1991.

Incompressible Navier-Stokes Equations in Cylindrical Polar Coordinates

The incompressible Navier-Stokes equations are given here in the (r, θ, z) coordinate system. The corresponding velocity components are, respectively (u, v, w).

$$\frac{1}{r}\frac{\partial}{\partial r}(ru) + \frac{1}{r}\frac{\partial v}{\partial \theta} + \frac{\partial w}{\partial z} = 0 \tag{A.18}$$

$$\rho\left(\frac{\partial u}{\partial t} + u\frac{\partial u}{\partial r} + \frac{v}{r}\frac{\partial u}{\partial \theta} - \frac{v^2}{r} + w\frac{\partial u}{\partial z}\right) = F_r - \frac{\partial p}{\partial r}$$
$$+\mu\left(\frac{\partial^2 u}{\partial r^2} + \frac{1}{r}\frac{\partial u}{\partial r} - \frac{u}{r^2} + \frac{1}{r^2}\frac{\partial^2 u}{\partial \theta^2} - \frac{2}{r^2}\frac{\partial v}{\partial \theta} + \frac{\partial^2 u}{\partial z^2}\right) \tag{A.19}$$

$$\rho\left(\frac{\partial v}{\partial t} + u\frac{\partial v}{\partial r} + \frac{v}{r}\frac{\partial v}{\partial \theta} + \frac{uv}{r} + w\frac{\partial v}{\partial z}\right) = F_\theta - \frac{1}{r}\frac{\partial p}{\partial \theta}$$
$$+\mu\left(\frac{\partial^2 v}{\partial r^2} + \frac{1}{r}\frac{\partial v}{\partial r} - \frac{v}{r^2} + \frac{1}{r^2}\frac{\partial^2 v}{\partial \theta^2} + \frac{2}{r^2}\frac{\partial u}{\partial \theta} + \frac{\partial^2 v}{\partial z^2}\right) \tag{A.20}$$

$$\rho\left(\frac{\partial w}{\partial t} + u\frac{\partial w}{\partial r} + \frac{v}{r}\frac{\partial w}{\partial \theta} + w\frac{\partial w}{\partial z}\right) = F_z - \frac{\partial p}{\partial z}$$
$$+\mu\left(\frac{\partial^2 w}{\partial r^2} + \frac{1}{r}\frac{\partial w}{\partial r} + \frac{1}{r^2}\frac{\partial^2 w}{\partial \theta^2} + \frac{\partial^2 w}{\partial z^2}\right) \tag{A.21}$$

The components of the stress tensor are:

$$\sigma_{rr} = -p + 2\mu \frac{\partial u}{\partial r}$$

$$\sigma_{\theta\theta} = -p + 2\mu \left(\frac{1}{r} \frac{\partial v}{\partial \theta} + \frac{u}{r} \right)$$

$$\sigma_{zz} = -p + 2\mu \frac{\partial w}{\partial z}$$

$$\tau_{r\theta} = \mu \left[r \frac{\partial}{\partial r} \left(\frac{u}{r} \right) + \frac{1}{r} \frac{\partial v}{\partial \theta} \right]$$

$$\tau_{\theta z} = \mu \left(\frac{\partial v}{\partial z} + \frac{1}{r} \frac{\partial w}{\partial \theta} \right)$$

$$\tau_{rz} = \mu \left(\frac{\partial u}{\partial z} + \frac{\partial w}{\partial r} \right)$$

Index